SpringerBriefs in Education

We are delighted to announce SpringerBriefs in Education, an innovative product type that combines elements of both journals and books. Briefs present concise summaries of cutting-edge research and practical applications in education. Featuring compact volumes of 50 to 125 pages, the SpringerBriefs in Education allow authors to present their ideas and readers to absorb them with a minimal time investment. Briefs are published as part of Springer's eBook Collection. In addition, Briefs are available for individual print and electronic purchase.

SpringerBriefs in Education cover a broad range of educational fields such as: Science Education, Higher Education, Educational Psychology, Assessment & Evaluation, Language Education, Mathematics Education, Educational Technology, Medical Education and Educational Policy.

SpringerBriefs typically offer an outlet for:

- An introduction to a (sub)field in education summarizing and giving an overview of theories, issues, core concepts and/or key literature in a particular field
- A timely report of state-of-the art analytical techniques and instruments in the field of educational research
- A presentation of core educational concepts
- An overview of a testing and evaluation method
- A snapshot of a hot or emerging topic or policy change
- An in-depth case study
- A literature review
- A report/review study of a survey
- An elaborated thesis

Both solicited and unsolicited manuscripts are considered for publication in the SpringerBriefs in Education series. Potential authors are warmly invited to complete and submit the Briefs Author Proposal form. All projects will be submitted to editorial review by editorial advisors.

SpringerBriefs are characterized by expedited production schedules with the aim for publication 8 to 12 weeks after acceptance and fast, global electronic dissemination through our online platform SpringerLink. The standard concise author contracts guarantee that:

- an individual ISBN is assigned to each manuscript
- each manuscript is copyrighted in the name of the author
- the author retains the right to post the pre-publication version on his/her website or that of his/her institution

More information about this series at http://www.springer.com/series/8914

Azra Moeed · Craig Rofe

Learning through School Science Investigation in an Indigenous School

Research into Practice

 Springer

Azra Moeed
School of Education
Victoria University of Wellington
Wellington, New Zealand

Craig Rofe
School of Education
Victoria University of Wellington
Wellington, New Zealand

ISSN 2211-1921   ISSN 2211-193X   (electronic)
SpringerBriefs in Education
ISBN 978-981-32-9610-7   ISBN 978-981-32-9611-4   (eBook)
https://doi.org/10.1007/978-981-32-9611-4

This Springer imprint is published by the registered company Springer Nature Singapore Pte Ltd.
The registered company address is: 152 Beach Road, #21-01/04 Gateway East, Singapore 189721, Singapore

Preface

Science investigation and practical work, in general, have been an important aspect of science education in most Western countries. Central to science is human curiosity and the need to understand the phenomena that unfold around us. To look for ways of explaining the world around us, we need an understanding that is persuasive, convincing, evidence based and open to critique.

The research reported in this book arose from a question we asked when conceptualising this project, what science investigations students in primary and secondary schools were carrying out and what they were learning from it. As well as how student learning through investigation could be enhanced (see Moeed & Anderson, 2018a, b). We were curious to know what science investigations Māori students (indigenous students) were carrying out and what they were learning from these investigations in Māori-medium schools? The New Zealand Curriculum requires that all students learn various approaches to investigating in their science education. The curriculum for Māori-medium schools is a translation of the English version and applies to Māori medium schools as well. About 15% of Māori students attend Māori medium schools, where the language of instruction is Te Reo Māori and the rest attend state schools where they are taught in English.

Here, we tell the story of how two teachers who did not have a background in science learnt and gained the confidence to teach science investigation and what the students learnt. When we approached a wharekura (school) and asked them to consider supporting our research, they were keen to collaborate. However, the two teachers chosen to participate told us that they did not know science and at the time no science was being taught at the kura. At this point, we could have said that the baseline data was not there and walked away but we asked the teachers how we could support them to teach science in Year 9? Together, we decided that the two researchers would organise ten 2-hour sessions where the teachers and students would learn together. The book sets out how this collaborative research worked and reports the outcomes for students and the teachers beyond the Year 9 class.

Briefly, it led to the development of a science programme in the kura to be taught in Te Reo Māori in the primary school and in English is the secondary school. We report the outcomes for the students in science and how they have been successful in national examinations for qualifications and have continued in science.

We also share our learnings and understanding of Mātauranga Māori (Māori knowledge), which our participating teachers considered to be inclusive of Western science. An exciting finding for us was that the students were drawing upon their Mātauranga when solving problems and how their engagement in investigation was much richer than keeping themselves limited to Western science.

The school has made some difficult decisions such as allowing science to be taught in English in senior school and we explain why they made these decisions. They believed that by Year 9 their students had developed a strong identity as Māori and were proficient in Te Reo Māori and they wanted the students to have the choice to learn science as well. For the researchers, we learnt about what a true partnership with Māori was about. It enabled us to learn how the students had both a science way and a Māori way of observing the natural world and explaining it. We note that Māori students' achievements in science are a concern for the country. This kura has given us an insight into how Māori students can be successful in science and continue in science. We share these insights in this book.

Azra Moeed is Indian but has lived and taught science for more than four decades in New Zealand at early childhood, primary, intermediate and secondary schools. Craig Rofe is from Ngāti Rangi, Te Atihaunui-ā-Pāpārangi. At present, he lectures in science education and Mātauranga Māori courses at Victoria University of Wellington. During the project, the researchers developed deeper understandings about the context, the teachers, the learners and Te Ao Māori.

Wellington, New Zealand   Azra Moeed
                                                                                            Craig Rofe

Acknowledgements We would like to thank The Teaching and Learning Initiative for funding the research project. We are grateful that Principal, teachers and students welcomed us in the school, gave us their time and shared their thoughts which made this book possible. We thank Dr. Abdul Moeed and Susan Kaiser for editorial support.

References

Moeed, A., & Anderson, D. (2018a). *Learning through school science investigation: Teachers putting research into practice*. Singapore: Springer.

Moeed, A., & Anderson, D. (2018b). Science investigation in secondary school. In *Learning through school science investigation* (pp. 71–91). Singapore: Springer.

Contents

Rārangi Kupu (Glossary)

Māori word	English translation
Aotearoa	New Zealand
kaiako	teacher
Kapa haka	Māori performing arts
kaupapa	cause, philosophy
KKM (Kura Kaupapa Māori)	Māori-medium schools based on Kura Kaupapa Māori philosophy
kōrero	talk, discourse
Kura	School
Kupu hou	New words
Māori	indigenous people of Aotearoa New Zealand
marae	Māori community centre, 'plaza'
Mātauranga Māori	Traditional Māori knowledge, Māori Education
Ngā manu kōrero	National Māori school speech competitions
Pūtaiao	Science in the Māori-medium curriculum
reo	Language, voice
Te Aho Matua	Title of legal definition of KKM
Te Kōhanga Reo	Māori immersion Early Childhood Education movement that preceded KKM
tikanga	Customs/ways of doing
tūpuna	Ancestors
tamariki	children

Chapter 1
Learning Through School Science Investigation

Knowledge is a *taonga* (treasure) handed down as '*taonga tuku iho*', that is, treasure from the ancestors, and as such is *tapu* (sacred). Knowledge is expressed in the form of personal power known as *mana*. How it is used is crucial (Bishop & Glynn 1999, p. 172).

1.1 Introduction and Background

New Zealand is a bicultural country and tries to give equal status to both the indigenous people of the land (Māori), and the Europeans (Pākehā) who settled in New Zealand after the Māori. Māori have their own language, which originally was only spoken, not written. Europeans brought English, which, through colonisation became the main written and spoken language for everyone. In the educational set-up, initially European knowledge, including scientific knowledge, became the primary discourse. However, for Māori, their ancestral knowledge, including scientific knowledge, remained with them as a valued part of their heritage. There is a lot already written about how Māori lived and how the arrival of Europeans brought change. For education, until the 1980s there was a state English-medium system for all students. However, Māori were able to initially negotiate and establish early childhood centres called Kohanga Reo (language nests) and later Māori-medium primary schools (kura) and secondary schools (wharekura). These schools are underpinned by Kaupapa Māori philosophy, which aims to revitalise the Māori language, Māori knowledge (Mātauranga) and cultural practices (Tikanga). The Māori-medium schools aspire to help students foster their identity as Māori.

This book reports research findings of a project that investigated the teaching and learning of Western science (Pūtaiao) in a wharekura in New Zealand and how it could be included in the school's curriculum. The focus of the research was what students learn through science investigation. Pūtaiao is often not included as a subject in wharekura due to the shortage of Māori-speaking teachers with science as an

© The Author(s), under exclusive licence to Springer Nature Singapore Pte Ltd. 2019
A. Moeed and C. Rofe, *Learning through School Science Investigation in an Indigenous School*, SpringerBriefs in Education, https://doi.org/10.1007/978-981-32-9611-4_1

undergraduate focus. Science is one of eight learning areas in *The New Zealand Curriculum* (Ministry of Education, 2007b); these are described in Sect. 1.3, Science Investigation and School Science Investigation.

This chapter backgrounds the context and information to situate the study in the wider context of Aotearoa New Zealand. The chapter begins with its unique bicultural past and present. We then turn to the issues of knowledge—Mātauranga and Pūtaiao. Next, we present a brief account of the current situation in relation to Māori students and their science education. In Māori education, culture matters, as do language and identity, so we explore these aspects in relation to Māori students' science learning.

We investigate cultural border-crossing, a notion theorised by Aikenhead (Aikenhead, 1996, 2001; Aikenhead & Ogawa, 2007) in relation to the indigenous people of Australia and discuss its importance in our context. Finally, as teacher beliefs have a major impact on what is valued, taught and how it is taught, we look at teachers' beliefs about science learning. All aspects are explored in relation to recent and extant literature. We acknowledge that we draw upon the scholarly works of Prof. Elizabeth McKinley and Assoc. Prof. Georgina Stewart, two leading scholars in the field of Māori science education (McKinley, 2005; McKinley & Stewart, 2009, 2012; Stewart, 2005, 2007, 2011, 2015).

Te Mātauranga o Aotearoa is the Māori translation of the *New Zealand Curriculum* (Ministry of Education, 2007a). Pūtaiao is used both as a general term for 'translated western science' and for traditional Māori knowledge (Stewart, 2011). Here, we use Pūtaiao as Western science only as taught in English and in Te Reo Māori (Māori language) in the research school.

The majority of students in New Zealand attend English-medium state schools; however, approximately 15% of Māori students attend wharekura. The wharekura's community, including the local iwi (tribe), hapū (sub-tribe) and whānau (family and wider family), can decide the attributes they want their students to develop during their schooling. Wharekura have a Kura Kaupapa Māori philosophy, which is culturally specific to Māori and aims to revitalise the Māori language, and Māori knowledge and culture (L. Smith, 2000). Kura Kaupapa Māori philosophy is based on theories of Māori knowledge, teaching and learning (Penetito, 2009; Sharples, 1994). The central belief is that Kura Kaupapa Māori has a positive impact on Māori students' educational achievement because it provides an environment in which Māori can enjoy educational success '*as* Māori' (Education Review Office, 2010) and retain cultural knowledge and identity while obtaining the Western equivalent. For details about the history of schooling and Māori schools, readers may be interested in the scholarly works of Wally Penetito, Graham Hingangaroa Smith and Georgina Stewart (Penetito, 2004; Smith, 2000, 2012; Stewart, 2012).

A wharekura has the responsibility for the physical, spiritual, mental and emotional development and teaching and learning needs of students, based on the principles of Te Ao Matua, a Māori designed curriculum.

The participating wharekura, where this research took place, is a co-educational, Year 1–13 school of about 250 at the time when the project began. Student numbers have continued to increase and were approximately 300 in the second year. All students identify their ethnicity as Māori (Education Review Office, 2016).

Students' holistic development as Māori had primacy as did the focus on students to develop an identity as Māori. A partnership with the kura required an understanding of the implicit ethical care of mana whenua (authority/power of place). For the researchers to collaborate with wharekura, having knowledge and respect for mana and knowledge of the wharekura and *how* to engage were important. The success of this research was dependent on the trust the wharekura, teachers and students placed in a broker (who was recommended by wharekura staff members), in this case, one of the researchers with Māori heritage. In addition to the broker's role, developing a relationship of care and support through home visits, being available when required and the notion of awhi mai, awhi atu (mutual support) gave the teachers confidence to establish a working relationship with experienced researchers and leadership in teaching science investigation. We collaborated with the principal and two teachers—Sue and Liz—and their classes. All names used in this book are pseudonyms.

1.1.1 *Mātauranga and Western Science*

Māori consider Mātauranga to be a taonga (treasure). Māori knowledge includes Māori ways of knowing that have been passed down through successive generations. The research presented here took place in a Māori-medium school, where Mātauranga Māori (Māori knowledge) includes Māori ways of knowing that have been passed down through successive generations. Broughton et al. (2015) argue that Mātauranga Māori is indigenous knowledge (IK) and has continued to expand through exploration and theorising at all levels of Māori society: 'It is a complete knowledge system that includes science' (Broughton et al., 2015, p. 88). Mercier and Leonard (2017) argue that 'IK is not a "thing" in the sense of a body of knowledge but active, complex, living and dynamic, and embedded in people and place' (p. 4). In the New Zealand context, IK is Mātauranga Māori.

Although complex, we attempt to explain how participants in this research saw knowledge, and in their view, what science is. Mātauranga Māori includes Pūtaiao/science and other ways of knowing, whereas Western science is not inclusive of indigenous knowledge, although some are beginning to accept it alongside Western science (Fig. 1.1).

> Educators are beginning to recognize that Western-based formal knowledge remains just one knowledge system of many. Though traditional knowledge has long been, and often continues to be, assigned a lower status in both development and scientific circles than Western-based science and technology, the value of IK (indigenous knowledge) in science has been receiving increasing attention. (Quigley, 2009, p. 75)

According to Stewart (2015), Science, by definition, is culture free; science claims to be universal knowledge, which applies equally everywhere (p. 401), whereas indigenous knowledge such as Mātauranga is embedded in the culture and passed on from generation to generation (McKinley, 2005). Tensions between these two kinds of

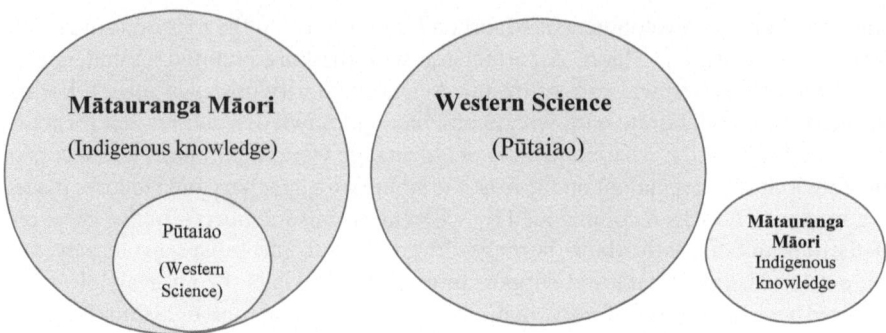

Fig. 1.1 Teachers' perceptions of the place of *school science* in the two knowledge systems

knowledge are well documented in the literature and are beyond the scope of this book (see, for example, McKinley, 2005; McKinley & Gan, 2014; McKinley & Stewart, 2009, 2012). The debate is broader and includes both ontological and epistemological issues in relation to indigenous knowledge, in Aotearoa New Zealand's case the tension is between Mātauranga Māori and Western science knowledge (McKinley, 2005; McKinley & Gan, 2014; McKinley & Stewart, 2009; Mercier & Leonard, 2017).

1.1.2 *Māori Science*

In our view, there can be two perspectives on what Māori science is depending on Western view or Māori view. The first would consider the collection of empirical data by Māori, the trends and patterns from this data as being Māori science. A Māori perspective could be that *any* way of making sense of the world, which is inclusive of a spiritual dimension, and Māori epistemology is also Māori Science. The issue of defining Māori science is a complex one. Here, we draw upon the works of Mercier and Leonard (2018). They raise the following question, *Is IK a science*? Then they suggest three possible answers. One, a view that indigenous people had no science, clearly not true, Māori, for example, did and continue to have their way of understanding the world which is evident in the advanced understandings of astronomy and navigation (Harris, Matamua, Smith, Kerr, & Waaka, 2013). Two, a view that there is no such thing as indigenous science (Science *is* Western). This view is strongly associated with the empirical nature of science. Some consider science to have the philosophical underpinning of inductive and deductive reasoning and for others, it is a method that includes a hypothesis, an experiment, generation of data and production of theory. This is a deficit view of Māori science. It implies that Māori did not explore their environment, or observe and collect data, or come up with their own theories about natural phenomena based on that data, which is Māori Science. The third view is probably the most interesting, that is 'indigenous knowledge is not and has never been Science, because it is superior to what is essentially an enterprise of political domination' (Mercier & Leonard, 2018, p. 5).

Expressing their views about science, Aikenhead and Ogawa (2007) state 'very purposefully and politically, the name *science* was chosen to replace *natural philosophy*' when the British Association for the Advancement of Science was established in 1831 (p. 542). The label *science* provided the founders of the society with a term that set them apart from natural philosophers and technologists.

We, as science education researchers, value both Mātauranga Māori and Pūtaiao and are interested in research that integrates the two fields of knowledge in a meaningful way.

1.1.3 Māori Students and Science Education

While at a philosophical level, scholars discuss the issues about science education for Māori students, recent research shows that Māori students are underperforming in science in both English-medium schools and wharekura, and that achievement in science is worse in wharekura (Stewart, 2011). Smith (1995) believed that Māori education underpinned by Kura Kaupapa Māori philosophy would help to overcome the disparity in achievement for Māori students; however, this has not been the case (Rofe, Moeed, Anderson, & Bartholomew, 2016; Stewart, 2011). Commenting on the continuing disparity in students' science achievement, Stewart (2007) attributes this lower achievement to the differences in worldviews between Māori and Pūtaiao, and a large amount of new science vocabulary (kupu hou) to be learnt in Te Reo Māori, a lack of teaching resources, and near absence of 'professional practice'. Providing for science learning is proving challenging for wharekura as there are few teachers who have both a background in science and who are also able to teach it in Te Reo Māori. Rofe et al. (2016) highlight the need for wharekura teachers to teach science while upholding Māori practice with an absence of a curriculum that reflects Māori epistemology. Various models have been tried, for example, teaching science in many wharekura through videoconferencing, but had limited use because of access to the Internet in kura, and having marae-based wānanga (immersion learning). The success of these programmes and the science that is taught are generally unknown as to date they have not been reported in the literature.

The issues about indigenous students' participation and achievement in science education have been highlighted and debated in science education literature for over three decades (McKinley & Gan 2014; Michie, Hogue, & Rioux, 2018). The focus internationally at present is on STEM (Science, Technology, Engineering, and Mathematics) education. Although the issues of science education for Māori students are well documented and research into indigenous methodologies and successful pedagogical approaches proposes ways in which to engage and support students (Bishop & Berryman, 2006; Bishop & Glynn, 1999), there has not been any significant improvement in student achievement or retention. Māori students do not find the STEM areas of education to be an attractive option for further education and there is a paucity of research that illuminates the funds of knowledge indigenous students bring to their classrooms (McKinley, 2016).

Science education plays a vital role in the gate-keeping function of schools in Western education. This is of overwhelming importance for our tamariki (children) and for Pūtaiao, yet it is hardly acknowledged or addressed in contemporary local education debates (Stewart, 2005). Within Te Ao Māori, educational success may be developing an understanding of whakapapa (genealogy), mōteatea (ancient lyrics) and karakia (incantations) for example, whereas in a Western approach, success in the senior sciences (biology, chemistry and physics) may be a marker of 'academic' success. Regardless of all the changes to our school qualifications, this Western view remains as true as ever (Elley, 2005). Few Māori students attending wharekura study science and those who do have poorer learning outcomes in sciences compared to the Māori students attending mainstream schooling due to the insufficient number of teachers having both a background in science and fluency in Te Reo Māori. This is a real matter of concern as all Māori aspire for their children to be successful in both.

Stewart (2011) analysed results of Māori students in science and mathematics and found that Māori students who were taking the National Certificate in Educational Achievement (the national qualification) could sit the examination in English or attempt a science paper which was written in English and then translated into Te Reo Māori. She reports that the achievement of students, whether they took the English or translated examination, was lower in Pūtaiao. She argues that the papers were not written for Pūtaiao; the examiners were translating Pūtaiao into Te Reo Māori but none of the content was underpinned by Mātauranga Māori. Stewart concludes that due to the educational policies and practices that supported the Māori-language goals, Pūtaiao exists as a dependent 'add-on' to science due to educational policy that requires its inclusion in the curriculum. Policies and practices in support of language status planning goals have, in this case, had the unintended effect of exacerbating underachievement in science education, while disallowing reform of the science curriculum according to the underlying principles of Kura Kaupapa Māori, and failing to support the goal of retaining traditional Māori knowledge.

1.1.4 Culture, Language, Identity

Classroom practice appears to present universalist views about science and the nature of science, and school curricula are resistant to change (McKinley, 2005). indigenous scholars have continually highlighted the need for school science to connect with students' cultural backgrounds (Cajete, 1995; Kawagley, 1995). This has been a challenge for mainstream science education where culturally responsive pedagogies have been promoted for science teachers to engage their Māori students. The Pākehā science teachers are either not engaging or are not able to include culturally responsive practices that are considered effective (Bishop & Berryman, 2009; Glynn, Cowie, Otrel-Cass, & Macfarlane, 2010). The connection that the Māori language has made with the renaissance of Māori culture in the past 50 years is not to be understated. Ambassadors of this renaissance, such as Sir James Henare, an eminent Māori who was instrumental in the Kohanga Reo, Māori-medium early childhood centres, described the Māori language as the life force of Māori mana 'Ko Te Reo te mauri o te mana Māori' (Waitangi Tribunal, 1989, p. 6). Similarly, McKinley (2005) states 'The essence of language is not only linguistic it is also the very existence of people. Hence it becomes an issue of identity for without it "difference" will become manifested in the physical attributes of a person' (p. 234). McKinley (2005) also argues that as languages carry the worldviews, the use of language and culture is important to Māori science education. She cites unexpected benefits of the revitalisation of the Māori language as it has become part of the English language and many New Zealanders know a significant number of Māori words. This has also created a unique place as we endeavour to celebrate the bicultural heritage of our nation. However, the creation of Māori words for scientific ideas has added complexity for the wharekura which need to attract not just teachers who have a background in science and can teach it through Te Reo Māori but also teachers who have the necessary kupu hou in Māori. The flow-on effect is that students need to learn the science, Te Reo Māori, and also the significant kupu hou (Stewart, 2011).

Identity defines who we are and our place in the world. For Māori adolescents, their racial-ethnic identity can be vital as it defines who they are and how they belong. Webber (2012) argues that for young Māori students, having a strong racial-ethnic identity may enable them to withstand the often-negative stereotypes that are present in society. A strong identity may also make them more resilient to negative external pressures and help them to engage and achieve in education. Opportunities to participate in, and have access to Te Ao Māori (The Māori world) are essential for the development of Māori students' identity as Māori (Faircloth, Hynds, Jacob, Green, & Thompson, 2016). McKinley, in a conversation about place and identity (Kincheloe, McKinley, Lim, & Barton, 2006), explains

> For me, developing a place-based education has mainly been about students knowing who they are, where they come from, to develop a sense of belonging. For many indigenous peoples, this concept of education is important because it simultaneously legitimates indigenous knowledges and acts as a mechanism to transfer knowledge intergenerationally. (p. 156)

Rofe et al. (2016) report that for the teachers in their research school, 'understanding Māori language and culture and developing identity as Māori were the first priority in their students' learning' (p. 6).

1.1.5 Cultural Border-Crossing

Māori students in mainstream schools often live and work in a culturally complex environment, where they need to negotiate a multitude of cultural nuances depending on the space. Science has a culture of its own (McKinley, 2016). Aikenhead (2006) suggests that teachers need to make the border-crossings their students have to make, explicit. He suggests that discourse and acknowledgement of students' prior knowledge and cultural ways of knowing would be helpful. Just as important is making it clear that there is border-crossing while teaching Western science ways of knowing, skills and values. Ocean Mercier a scholar who is the first Māori woman with a doctorate in Physics talks about her turbulent border-crossings from being a Physics academic to becoming a Māori academic. Talking about her academic migration Mercier (2014) says

> And finally, experiencing the uncertainty at the fault-line of disciplinary movements, and engaging new problems with tools from both sides is stimulating and fun. Rather than remaining entrenched in one side or the other, migration allows us to work the fertile land at the disciplinary faultline (p. 75).

For the students in the wharekura, this is particularly important as they have been situated within a Māori ontology for all of their lives and the orientation with a subject firmly placed within a Western paradigm is challenging.

1.1.6 Teacher Beliefs and Teaching Science

Teacher beliefs about science teaching and learning are influential factors in the development of pedagogical content knowledge for science teaching and a critical factor in the implemented curriculum (Anderson, 2015). Teachers' beliefs about the goals and purposes of science teaching and their understanding of the nature of science also affect their practice (Friedrichsen, Van Driel, & Abell, 2011). Mansour (2009) argues that teachers' thinking is influenced by past experiences and is context specific. Further, Mansour (2013) highlights the consistencies and inconsistencies between teacher beliefs and practices. Historically, scholars have emphasised a strong relationship between teacher knowledge and beliefs (Shulman, 1987), and Calderhead (1996) argued that knowledge is 'the factual propositions and the understandings that inform skilful action' and beliefs are 'generally referring to suppositions, commitments, and ideologies' (p. 715). Kleickmann, Tröbst, Jonen, Vehmeyer, and Möller (2016) investigated teacher beliefs, motivations, instructional practices and student achievement. They found that additional expert scaffolding during curriculum-specific professional development based on curriculum materials is very useful for the preparation of elementary teachers for teaching science. Rofe et al. (2015) have found that teacher confidence in their science content knowledge and knowledge of science were barriers that wharekura teachers had to overcome to teach science. As teachers' beliefs and confidence affect their actions, they are relevant to the research presented here.

1.1.7 Researchers

The researchers had been science teachers and are currently science teacher educators. One researcher is Indian but has lived for more than four decades and taught science in New Zealand at early childhood, primary, intermediate and secondary schools in New Zealand (Author one, Azra). The other researcher is Māori and at present lectures in science education and teaches Matauranga Māori courses at the university (Author two, Craig). During the project, the researchers developed deeper understandings about the context, the teachers, the learners and Te Ao Māori (details in Chap. 7).

1.1.8 Enabling Leadership

Educational leadership plays a critical role in improving educational outcomes for indigenous students (Hohepa, 2013). When faced with the challenge of not being able to attract teachers who had a background in science and who could teach it through Te Reo Māori, the wharekura principal and whānau made a difficult decision to allow for Pūtaiao professional development of teachers to be delivered in English. This decision was a major compromise and would lead to a change which was contrary to the Kura philosophy of teaching in a fully immersed Te Reo Māori environment. Wharekura can decide on the attributes they want their students to have when they leave school. There was a recognition that their students were already developing a strong identity as Māori, learning about Mātauranga Māori and learning Te Reo Māori. They wanted their students to have access to Western science and the choice of learning it in the senior school and beyond and to experience success in it. The leadership took an important step; to implement this they started with two teachers who did not have a background in science and by participating in this research project could create a space for science in their school. Their commitment is reflected in the wharekura, not only by having a science programme for Years 1–10 but also by employing science graduates who could teach science in English in the senior school. The detailed methodology is presented in the following chapter.

1.2 The New Zealand Curriculum, Science and Science Investigation

The *New Zealand Curriculum* (Ministry of Education, 2007b) has eight learning areas. The science learning area aims for students to develop conceptual, procedural and epistemological understandings of science during their schooling and prioritises scientific literacy. The curriculum has four contextual strands: The Living world;

Physical world; Material world; and Planet Earth and Beyond. Importantly, the curriculum has an overarching Nature of Science strand which aims for students to develop:

Understanding about science Learn about science as a knowledge system: the features of scientific knowledge and the processes by which it is developed; and learn about the ways in which the work of scientists interacts with society.

Investigating in science Carry out science investigations using a variety of approaches: classifying and identifying, pattern-seeking, exploring, investigating models, fair testing, making things or developing systems.

Communicating in science Develop knowledge of the vocabulary, numeric and symbol systems and conventions of science and use this knowledge to communicate about their own and others' ideas.

Participating and contributing Bring a scientific perspective to decisions and actions as appropriate. (http://nzcurriculum.tki.org.nz/The-New-Zealand-Curriculum, p. 23)

Pūtaiao i roto i te Mātauranga o Aotearoa. (Ministry of Education, 2007a), as noted, is the Māori translation of the *New Zealand Curriculum* (Ministry of Education, 2007b) and thus requires students who attend Māori-medium schools to learn about the nature of Pūtaiao. Although there is considerable knowledge about New Zealand students' science learning in mainstream education, due to a paucity of research little is known about the Pūtaiao/science Māori students are learning in the wharekura.

The Pūtaiao curriculum (Ministry of Education, 2007a) views knowledge as 'the two repositories/systems (metaphorically: baskets)' (p. 35). This view accommodates Māori and Western knowledge relating to the natural world (*Te Tāhuhu o Te Mātauranga*/Translation of Te Mārautanga o Aotearoa: Hei tauira hei korerorero; Ministry of Education, 2007a). According to this curriculum, 'knowledge is a human product against the sensory experience of the world; it is flexible, fallible knowledge, which is subject to continual review' (p. 35). The intention of Pūtaiao is to enable the Māori world to face and embrace the future.

Similar to Science in the *New Zealand Curriculum,* the Pūtaiao document has contextual strands: the Natural World; the Physical World; and the Material World. Whereas Planet Earth and Beyond is a separate strand in science, it is integrated within the Natural World in the Pūtaiao document.

The Nature of Pūtaiao
This strand (Ministry of Education, 2007a) envelops and underpins all of the others. There are five parts:

UNDERSTANDING: Learn about the knowledge system of science: the role of central concepts and theories; how it develops and advances; and its relationships to the individual, to society, and to the environment. Develop critical abilities in order to carefully evaluate the world of science from a Māori viewpoint.

SKILLS: Complete various types of investigation: classifying and identifying; exploration; pattern-seeking; fair testing; problem-solving; making a product or a model.

VALUES: Develop an increasing understanding of: cognitive and moral values; ancient and modern narratives; and the philosophies of science. Hold fast to traditional values originating from a notion of the relationships between all living things.

COMMUNICATING IDEAS: Develop competence in scientific literacy, numeracy and the symbol systems of science, and use these literacies to communicate their own ideas, and the ideas of others.

SOCIAL ACTION: Relate their knowledge of science to social decisions and actions: to take care of the environment in the vicinity of their home, and to think about issues impacting on their iwi, and wider issues, relating to the environment. (p. 38)

In its present form, students are expected to learn about the nature of science from a Māori perspective. They are expected to become skilful in conducting various types of investigation and become scientifically literate. As can be seen, there are differences between the requirements of the Nature of Science and the Nature of Pūtaiao strands with the latter including values and social action which also reflects the Kura Kaupapa Māori philosophy.

1.3 Science Investigation and School Science Investigation

Scientific knowledge is created when researchers ask a question and investigate to find an answer to that question, so the purpose of science investigation is empirical. Scientists have considerable knowledge and understand the nature of their discipline. They also have skills and procedural understanding to draw upon when they investigate. School science is different; here the purpose is to help students understand science ideas and theories, develop investigative skills, begin to understand how science works and become scientifically literate (Moeed & Anderson, 2018). Being a scientifically literate citizen is becoming increasingly important for students so that they can make informed decisions about everyday choices they make in their personal lives as well as the impact of their choices on the planet we inhabit.

Before we delve into the details, some clarification is needed in terms of the various terminologies used when talking about school science investigations. In some countries, for example, the United States, science investigation is called science inquiry. Then there are terms that are used interchangeably in school science and science education literature. These include practical work, experiments and investigations. The readers may find the following definitions useful:

Practical Work
Practical work is any science teaching and learning activity, where students observe or manipulate objects individually or in small groups. Practical work has generally been called hands-on activities or doing science; it may or may not take place in the laboratory (Millar, 2004a, b).

Experiments

An experiment is a specific form of investigation where intervention is performed to create a phenomenon that can be observed either quantitatively (by measuring) or qualitatively. Experiments are different from investigations in which data are collected by observation or something that is taking place (Hacking, 1983). The key criterion is 'whether you have to do something to create a phenomenon you are interested in and hence generate data, or can simply observe what is happening, or what exists, anyhow' (Millar, 2015, p. 418).

School Science Investigation

A school science investigation is an activity requiring identification of a question, using both conceptual and procedural knowledge in planning and carrying out the investigation, gathering, processing and interpreting data, and drawing conclusions based on evidence. Ideally, the process is iterative and the student has some choice in what they want to investigate (Millar, 2010, p. 108).

(For further detail, see Abrahams & Millar, 2008; Moeed & Anderson, 2018.)

1.3.1 Similarities and Differences Between Scientists' Investigation and School Science Investigation

As shown in Fig. 1.2, there are similarities between scientists' investigation and

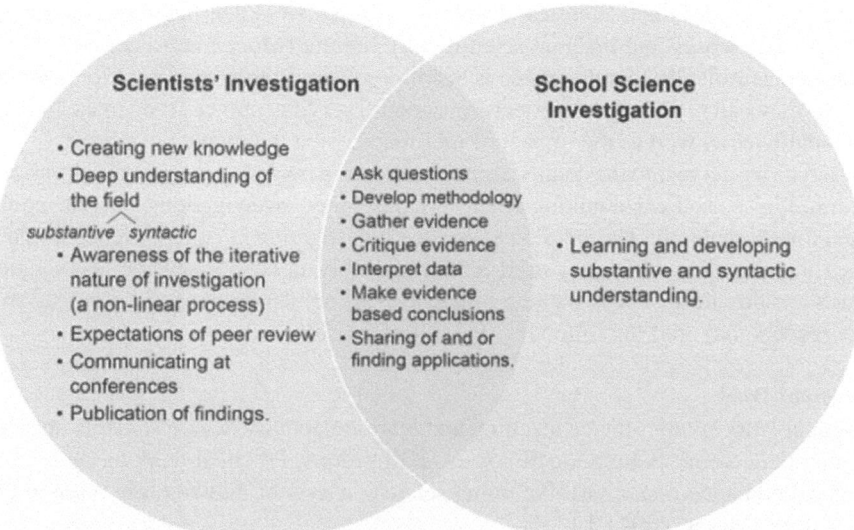

Fig. 1.2 Similarities and differences between scientists' science investigation and school science investigation (Moeed & Anderson, 2018, p. 11)

school science investigations, as the purposes for both are different. As already noted, scientists' investigations are empirical and endeavour to generate new knowledge, whereas the purpose of school science investigation is to develop substantive and syntactic understandings. The term *substantive* is used to describe the body of knowledge generated by a discipline, which in science is concerned with understanding the physical, material and natural worlds, whereas syntactic knowledge is about understanding the processes of knowledge creation and acceptance (Moeed & Anderson, 2018; Schwab, 1964). Both scientists and students ask questions, develop methodology, gather evidence, critique evidence, interpret data, make evidence-based conclusions and share and apply their findings, both doing so at their own level. This area of syntactic knowledge has been termed the Nature of Science and is a leading focus for the professional development of New Zealand Science teachers (Moeed & Anderson, 2018).

Internationally, the number of students opting to study science in secondary schools is declining, although there is general agreement that all students ought to learn science at least up to junior high school, and possibly beyond (Osborne, Simon, & Collins, 2003). In New Zealand, science is compulsory up to Year 10 (age 14). A disproportionately few Māori students continue with science through to Year 13 (age 17), the final year of school (Rata, 2012). Recent results of the Organization for Economic Co-operation and Development Programme for International Student Assessment for scientific literacy show that although the achievement of New Zealand students is generally high, Māori students are underperforming (Glynn et al. 2010). A similar pattern is seen in the National Education Monitoring Project for science where Māori students are, on average, achieving lower scores than non-Māori students (Woods-McConney et al. 2013).

Increasing student numbers in STEM subjects is both a national and international priority. Raising Māori student achievement in general, and in science in particular, is also a government priority. The number of Māori students taking STEM subjects is proportionately lower than the remaining student population taking them:

Ko ngā pae tawhiti whāia kia tata, ko ngā pae tata whakamaua

The potential for tomorrow depends on what we do today. (Ministry of Education, statement of intent for 2014–18[1])

However, what we do tomorrow needs to be based on knowledge about the pūtaiao/science Māori students are learning in wharekura today. Without such knowledge, we lack the foundation on which to improve outcomes for Māori students and for New Zealand.

This book reports a case study of science investigation in a Māori-medium school, the related teacher professional learning, and student outcomes.

---

[1] (https://www.education.govt.nz/assets/Documents/Ministry/Publications/Statements-of-intent/2014SOI.pdf).

References

Abrahams, I., & Millar, R. (2008). Does practical work really work? A study of the effectiveness of practical work as a teaching and learning method in school science. *International Journal of Science Education, 30*(14), 1945–1969. https://doi.org/10.1080/09500690701749305.

Aikenhead, G. S. (1996). Science education: Border crossing into the subculture of science. *Studies in Science Education, 27*(1), 1–52. https://doi.org/10.1080/03057269608560077.

Aikenhead, G. (2001). Integrating western and Aboriginal sciences: Cross-cultural science teaching. *Research in Science Education, 31*(3), 337–355. https://doi.org/10.1023/a:1013151709605.

Aikenhead, G. S. (2006). *Science education for everyday life: Evidence-based practice*. New York: Teachers College Press.

Aikenhead, G. S., & Ogawa, M. (2007). Indigenous knowledge and science revisited. *Cultural Studies of Science Education, 2*(3), 539–620. https://doi.org/10.1007/s11422-007-9067-8.

Anderson, D. (2015). The nature and influence of teacher beliefs and knowledge on the science teaching practice of three generalist New Zealand primary teachers. *Research in Science Education, 45*(3), 395–423. https://doi.org/10.1007/s11165-014-9428-8.

Bishop, R., & Berryman, M. (2006). *Culture speaks: Cultural relationships and classroom learning*. Wellington: Huia.

Bishop, R., & Berryman, M. (2009). The Te Kotahitanga effective teaching profile. *Set: Research Information for Teachers, 2*(2), 27–33.

Bishop, R., & Glynn, T. (1999). Researching in Māori contexts: An interpretation of participatory consciousness. *Journal of Intercultural Studies, 20*(2), 167–182. https://doi.org/10.1080/07256868.1999.9963478.

Broughton, D., Te Aitanga-a-Hauiti, T., Porou, N., McBreen, K., Waitaha, K. M., & Tahu, N. (2015). Mātauranga Māori, tino rangatiratanga and the future of New Zealand science. *Journal of the Royal Society of New Zealand, 45*(2), 83–88. https://doi.org/10.1080/03036758.2015.1011171.

Cajete, G. (1995). *Look to the mountain: An ecology of indigenous education*. Durango, CO: Kivaki Press.

Calderhead, J. (1996). Teachers: Beliefs and knowledge. In D. C. Berliner & R. C. Calfee (Eds.), *Handbook of educational psychology* (pp. 709–725). New York: Simon & Schuster Macmillan.

Education Review Office. (2010). *Promoting success for Māori students: Schools' progress (June 2010)*. Retrieved from www.ero.govt.nz/National-Reports/Promoting-Successfor-Maori-Students-Schools-Progress-June-2010.

Education Review Office. (2016). *School report*. Retrieved from https://www.ero.govt.nz/review-reports/te-kura-maori-o-porirua-26-09-2016/#2-uri-outcomes.

Elley, W. (2005). On the remarkable stability of student achievement standards over time. *New Zealand Journal of Educational Studies, 40*(1–2), 3–23.

Faircloth, S. C., Hynds, A., Jacob, H., Green, C., & Thompson, P. (2016). Ko wai Au? Who am I? Examining the multiple identities of Māori youth. *International Journal of Qualitative Studies in Education, 29*(3), 359–380. https://doi.org/10.1080/09518398.2015.1053158.

Friedrichsen, P., Van Driel, J. H., & Abell, S. K. (2011). Taking a closer look at science teaching orientations. *Science Education, 95*(2), 358–376. https://doi.org/10.1002/sce.20428.

Glynn, T., Cowie, B., Otrel-Cass, K., & Macfarlane, A. (2010). Culturally responsive pedagogy: Connecting New Zealand teachers of science with their Māori students. *The Australian Journal of Indigenous Education, 39*(1), 118. https://doi.org/10.1375/s1326011100000971.

Hacking I. (1983). *Representing and intervening*. Cambridge: Cambridge University Press.

Harris, P., Matamua, R., Smith, T., Kerr, H., & Waaka, T. (2013). A review of Māori astronomy in Aotearoa-New Zealand. *Journal of Astronomical History and Heritage, 16*(3), 325–336.

Hohepa, M. K. (2013). Educational leadership and indigeneity: Doing things the same, differently. *American Journal of Education, 119*(4), 617–631. https://doi.org/10.1086/670964.

Kawagley, A. O. (1995). *A Yupiak worldview: A pathway to ecology and spirit*. Prospect Heights, IL: Waveland Press.

Kincheloe, J. L., McKinley, E., Lim, M., & Barton, A. C. (2006). A conversation on 'sense of place' in science learning. *Cultural Studies of Science Education, 1*(1), 143–160. https://doi.org/10.1007/s11422-005-9003-8.

Kleickmann, T., Tröbst, S., Jonen, A., Vehmeyer, J., & Möller, K. (2016). The effects of expert scaffolding in elementary science professional development on teachers' beliefs and motivations, instructional practices, and student achievement. *Journal of Educational Psychology, 108*(1), 21. https://doi.org/10.1037/edu0000041.

Mansour, N. (2009). Science teachers' beliefs and practices: Issues, implications and research agenda. *International Journal of Environmental and Science Education, 4*(1), 25–48.

Mansour, N. (2013). Consistencies and inconsistencies between science teachers' beliefs and practices. *International Journal of Science Education, 35*(7), 1230–1275. https://doi.org/10.1080/09500693.2012.743196.

McKinley, E. (2005). Locating the global: Culture, language and science education for indigenous students. *International Journal of Science Education, 27*(2), 227–241. https://doi.org/10.1080/0950069042000325861.

McKinley, E. (2016, August). Melbourne: STEM and indigenous learners. Presentation at Australian council for education research.

McKinley, E., & Gan, M. J. (2014). Culturally responsive science education for indigenous and ethnic minority students. In S. K. Abell & N. G. Lederman (Eds.), *Handbook of research on science education* (2nd ed., pp. 284–300). Mahwah, NJ: Lawrence Erlbaum.

McKinley, E., & Stewart, G. M. (2009). Falling into place: Indigenous science education research in the Pacific. In S. Ritchie (Ed.), *World of science education* (Vol. 2, pp. 49–66)., Handbook of research in Australasia Rotterdam: Sense.

McKinley, E., & Stewart, G. M. (2012). Out of place: Indigenous knowledge (IK) in the science curriculum. In B. Fraser, C. McRobbie, & K. Tobin (Eds.), *Second international handbook of science education* (pp. 541–554). New York: Springer.

Mercier, O. R. (2014). At the Faultline of Disciplinary Boundaries: Emigrating from Physics to Māori Studies. In C. Mason & F. Rawlings-Sanaei (Eds.), *Academic migration, discipline knowledge and pedagogical practice*. Singapore: Springer.

Mercier, O. R., & Leonard, B. (2017). Indigenous knowledge(s) and the sciences in global contexts: Bringing worlds together. *Handbook of Indigenous Education*, 1–29.

Michie, M., Hogue, M., & Rioux, J. (2018). The application of both-ways and two-eyed seeing pedagogy: Reflections on engaging and teaching science to post-secondary indigenous students. *Research in Science Education, 48*(6), 1205–1220. https://doi.org/10.1007/s1116.

Millar, R. (2004a). *The role of practical work in the teaching and learning of science* (pp. 1–24). High school science laboratories: Role and vision.

Millar, R. (2004b). *The role of practical work in the teaching and learning of science*. Paper prepared for the meeting High school science laboratories: Role and vision. Washington, DC: National Academy of Sciences.

Millar, R. (2010). Practical work. In J. Osborne & J. Dillon (Eds.), *Good practice in science teaching: What research has to say* (2nd ed., pp. 108–134). Maidenhead, UK: Open University Press.

Millar, R. (2015). Experiments. In R. Gunstone (Ed.), *Encyclopaedia of science education* (pp. 418–419). Dordrecht: Springer.

Ministry of Education. (2007a). *Pūtaiao i roto i te Marautanga o Aotearoa*. Wellington: Learning Media.

Ministry of Education. (2007b). *The New Zealand curriculum*. Wellington: Learning Media.

Moeed, A., & Anderson, D. (2018). *Learning through school science investigation: Teachers putting research into practice*. Singapore: Springer.

Osborne, J., Simon, S., & Collins, S. (2003). Attitudes towards science: A review of the literature and its implications. *International Journal of Science Education, 25*(9), 1049–1079. https://doi.org/10.1080/0950069032000032199.

Penetito, W. (2004). Research and context for a theory of Maori schooling. *Intercultural Communication, 173*–188.

Penetito, W. (2009). Place-based education: Catering for curriculum, culture and community. *New Zealand Annual Review of Education, 18*(2008), 5–29.

Quigley, C. (2009). Globalization and science education: The implications for indigenous knowledge systems. *International Education Studies, 2*(1), 76–88.

Rata, E. (2012). Theoretical claims and empirical evidence in Māori education discourse. *Educational Philosophy and Theory, 44*(10), 1060–1072. https://doi.org/10.1111/j.1469-5812.2011.00755.x.

Rofe, C., Moeed, A., Anderson, D., & Bartholomew, R. (2016). Science in an indigenous school: Insight into teacher beliefs about science inquiry and their development as science teachers. *The Australian Journal of Indigenous Education, 45*(1), 91–99. https://doi.org/10.1017/jie.2015.32.

Schwab, J. J. (1964). The structure of the disciplines: meanings and significances. In G. W. Ford, & L. Pugno (Eds.), *The structure of knowledge and the curriculum* (pp. 6e30). Chicago: Rand McNally.

Sharples, P. (1994). Kura kaupapa Māori. In H. McQueen (Ed.), *Education is change* (pp. 11–21). Wellington: Bridget Williams.

Shulman, L. (1987). Knowledge and teaching: Foundations of the new reform. *Harvard Educational Review, 57*(1), 1–23.

Smith, L. T. (1995). *Kaupapa Maori research*. Paper presented at Te Matawhanui Hui, Massey University, Palmerston North.

Smith, G. H. (2000). Maori education: Revolution and transformative action. *Canadian Journal of Native Education, 24*(1), 57.

Smith, L. T. (2000). *Kaupapa Maori principles and practices: A literature review*. Palmerston North: International Research Institute for Maori and Indigenous Education, Massey University.

Smith, G. H. (2012). The politics of reforming Maori education: The transforming potential of Kura Kaupapa Maori. *Towards Successful Schooling, 73*–87.

Stewart, G. (2005). Māori in the science curriculum: Developments and possibilities. *Educational Philosophy and Theory, 37*(6), 851–870. https://doi.org/10.1111/j.1469-5812.2005.00162.x.

Stewart, G. M. (2007). *Kaupapa Māori science* (Doctoral diss.). The University of Waikato, Hamilton, New Zealand.

Stewart, G. (2011). Science in the Māori-medium curriculum: Assessment of policy outcomes in Pūtaiao education. *Educational Philosophy and Theory, 43*(7), 724–741. https://doi.org/10.1111/j.1469-5812.2009.00557.x.

Stewart, G. M. (2012). Achievements, orthodoxies and science in Kaupapa Maori schooling. *New Zealand Journal of Educational Studies, 47*(2), 51.

Stewart, G. M. (2015). Ethnoscience (pp. 401–402). *Encyclopedia of science education*. Netherlands: Springer.

Waitangi Tribunal. (1989). *Report of the Waitangi Tribunal on the Te Reo Māori claim. Wai 11*. Wellington: Author.

Webber, M. (2012). Identity matters: Racial-ethnic identity and Maori students. *Set: Research Information for Teachers, 2*, 20.

Woods-McConney, A., Oliver, M. C., McConney, A., Maor, D., & Schibeci, R. (2013). Science engagement and literacy: A retrospective analysis for indigenous and non-indigenous students in Aotearoa New Zealand and Australia. *Research in Science Education, 43*(1), 233–252. https://doi.org/10.1007/s11165-011-9265-y.

...

Chapter 2
Research Design and Methodology

2.1 Introduction

Teaching and learning of science in New Zealand primary schools is an ongoing concern. As mentioned in the previous chapter, Pūtaiao/Science was not being taught at the research school, which was a Māori-medium school where Te Reo Māori was used as a medium of instruction. The research design was, therefore, tailored to the needs of the research school as explained below. A brief justification of the research design is presented here and is supported by qualitative research and case study literature. Research evidence suggests that teachers' perceptions about the purposes of learning science vary among primary and secondary schools. While it is known that teachers have multiple goals for learning through science investigation, little is known about teaching and learning to investigate in wharekura. Internationally, research over the past three decades has found little evidence that students learn what the teacher intends them to learn by participating in practical work, which includes investigations (Abrahams & Millar, 2008; Abrahams & Reiss, 2012; Millar, 2011, 2012).

Abrahams (2009) suggests that teachers do investigative work because they say it is motivational, and students consider it to be a more attractive alternative to copying notes. In a case study in New Zealand, it is reported that when teachers had clear modest goals for the intended learning and shared these with the students, the students did learn from practical investigations (Moeed & Anderson, 2018). Little is known about what Pūtaiao/science students are being taught in wharekura and if any science investigations were carried out. Hence to explore this, it became the focus of our research.

© The Author(s), under exclusive licence to Springer Nature Singapore Pte Ltd. 2019   19
A. Moeed and C. Rofe, *Learning through School Science Investigation in an Indigenous School*, SpringerBriefs in Education, https://doi.org/10.1007/978-981-32-9611-4_2

2.2 Research Questions

We set out to investigate what participating teachers, with no background in science, believed was the reason their students should be learning science, what investigating in science is and how they could teach it. We had the following research questions.

1. Within school science, how do participating teachers conceptualise science investigation and its educational role?
   Sub-questions:

   a. *What purposes do they see for school science and science investigation?*
   b. *What are teachers' beliefs about science and science investigations?*
   c. *How do the teachers manage the tension between Mātauranga Māori while teaching Western science?*

2. How do participating teachers practise science and what can they do to enhance student learning through science investigation?
   Sub-questions:

   a. *How do participating teachers teach science and science investigation?*
   b. *What approaches to science investigation do they practise?*

3. What do the students learn from participating in science and science investigation?

2.3 The Theoretical Frame

Wharekura are based on Kura Kaupapa Māori philosophy, culturally specific to Māori, with the aim of ensuring the survival of the Māori language, knowledge, and culture (Smith, 2000). Kura Kaupapa Māori has developed from theories based on Māori knowledge and traditional Māori ways of teaching and learning (Penetito, 2009; Sharples, 1994). The fundamental belief is that Kura Kaupapa Māori provides an environment in which Māori can enjoy educational success *as* Māori (ontologically) and it has a positive impact on Māori students' educational achievement (Education Review Office, 2013). As one of the two researchers was non-Māori and Pūtaiao teaching during the research was in English, it was not appropriate to use Kaupapa Māori as a theoretical frame. The theoretical frame of this research was a social constructivist lens.

Much of science education research has been underpinned by a constructivist theory of learning. The central constructivist position is that new knowledge is constructed by a learner based on what they already know. Constructivism gives priority to the learner's prior knowledge (Driver, Asoko, Leach, Scott, & Mortimer, 1994). Taber (2016) provides a useful visual of the many ways in which constructivism has been used including: as a philosophical position; in research tradition, as a principle of cognitive development; and as a perspective on teaching and learning. Taber views constructivism in science education as '**a perspective on learning that has consequences for how to teach canonical knowledge**' (p. 117, the emphasis is original). In this research, and based on Taber's classification, we have used constructivism as a lens for three different purposes:

1. For developing and evaluating pedagogy, teaching choices made by the participants, e.g. teaching sequence and models, metaphors and analogies, learning activities
2. To explore teacher and student thinking as a starting point and how it develops and
3. To view the social context which includes institutional, class, curriculum and cultural contexts.

Our view is that knowledge is constructed by the learner but is socially mediated. What we mean by this is that the learner learns from social tools, which include digital media and books, as well as by interaction with the teacher and other students. Our lens, therefore, is social constructivist (Solomon, 1987). As the theoretical lens determines the research design we agreed on a qualitative, interpretive paradigm for our study.

2.4 Research Design and Methodology

Within a qualitative paradigm, we considered several methodological approaches including ethnography, narrative, phenomenology, grounded theory and case study. As we were interested in gaining a deep understanding of Pūtaiao teaching and learning, we decided that a case study of science teaching and learning in a Māori-medium school was most appropriate. Interpretive research aims to communicate to its audience the understandings developed by observing and recording the everyday life of the participants (Merriam, 2001; Patton, 2002; Stake, 1995). Grbich (2007) suggest, 'Multiple realities are presumed, with different people experiencing these differently' (p. 8). Our intention was to gain an in-depth and comprehensive picture of a variety of students' and teachers' experiences and perceptions of science investigation and these were reflected in our choice of methodology and data gathering tools. Although case studies that provide a detailed and in-depth understanding of the multiple variables of the classroom and that describe the student–teacher interactions have increased, few have investigated science teaching and learning in a wharekura. Case studies are useful to enhance our understandings of science learning in classrooms. Cases have boundaries, revealing 'how all the parts work together to form a whole' (Merriam, 2001, p. 6). Our case study was bounded by the level of schooling (in this case, Years 9 and 10), place (the kura and the designated classroom in which the teaching occurred), and the participants, the two teachers and their classes. Only science classes were observed so the case was also bounded by the learning area. Yin (1994) describes 'a case study as an empirical inquiry that investigates a contemporary phenomenon within its real life context, especially when the boundaries between phenomenon and context are not clearly evident' (p. 13). However, it did involve the participants learning about science and science investigations through a different perspective, Māori and Western science.

A case study provides a number of options. Stake (2005) suggests that according to purposes for studying cases there are three types of case studies—'Intrinsic case study where a researcher wants to better understand a particular case, Instrumental case studies that provide insights into an issue or refine a theoretical explanation and Collective case studies that involve extensive study of several instrumental cases' (p. 378). The present research was an intrinsic case study as our interest was to understand the phenomenon of interest, teaching and learning of science investigation in this context.

The participants were two teachers, Sue and Liz (pseudonyms), and their class of six girls and eight boys. It was a combined class of Year 9 and 10 students (age 13–15). Sue was a trained early childhood and primary school teacher and had over 17 years' teaching experience at the kura. She was fluent in te reo Māori, and had considerable cultural knowledge as she was of Māori descent. Liz was a trained high school social sciences teacher in her fifth year of teaching and also had Māori whakapapa. She had taught English, information and communication technology (ICT), and statistics in te reo Māori at this kura for 3 years. Both teachers were competent ICT users.

A variety of data gathering tools were used including teacher interviews, teacher-stimulated response interviews, classroom observations, student focus group interviews and student questionnaires.

The following data sources were used during the 2 years of this research:

- A semi-structured interview with both teachers to explore their views about the purposes of science, and their beliefs about why and how students should learn Pūtaiao.
- The unit plans pertaining to the topic taught during data collection.
- Classroom observations of workshops conducted by one researcher, where the other researcher was teaching (pre-phase) (see Appendix 2.1).
- Classroom observations of three lessons for each class where teachers had implemented an investigative approach. The lessons were audio-recorded, and a running record was kept and field notes were taken to gain insight into learning opportunities (Phase 1 and Phase 2).
- Following each observed lesson, a focus group interview conducted with five/six students selected by the teachers. Interviews inquired into students' perceptions about their substantive and syntactic learning and their perceptions of the purpose and role of school science investigation.
- Artefacts resulting from the research investigation including student work, short video-clips, and audio-recordings of conversation with students collected as appropriate during the investigation.
- Teacher and researcher reflections at the end of the data collection.

The research team coded the data inductively. Emerging codes and themes were discussed, and examples and exemptions cross-checked using a process of constant comparison (Merriam, 2001). The codes were generated by one researcher and cross-checked by the other researcher. Accuracy and common understanding were required

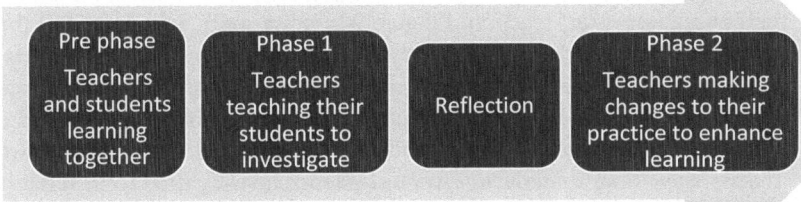

Fig. 2.1 Data collection in pre-phase, Phases 1 and 2

for consistency in allocating the codes. A constant comparison process (Merriam, 1998) was used in open coding of phrases and words used by participants.

Data collection took place in three phases, as illustrated in Fig. 2.1.

2.5 Ako (Reciprocal Learning)

The Māori notion of ako views learning as a reciprocal process ('ako' is the word for both teaching and learning). The research design accommodated a collaborative approach, where researchers had the opportunity to learn about the context, pedagogies and participants' approaches to teaching, learning and Māori ontology. At the same time, the teachers, with guidance from the researchers, learnt and taught science including teachers learning how to plan and carry out science investigations and co-constructing science investigations with their students. This collaborative approach was also a priority for the Teaching and Learning Research Initiative (TLRI) who funded the research.

At the start of this research, the participating wharekura did not have a science programme as it was difficult to employ teachers who could teach science in Te Reo Māori. Consequently, students with whom we worked had no formal science learning. At this point, in the interest of meaningful collaboration, we did not want to say that the project could not go ahead and walk away from it. Instead, we asked the participating teachers what would help them to teach science. They said that they did not have a background in science and did not feel confident to teach it. This led to a change in design for the case study. The research was one of the three case studies in our larger research project involving mainstream primary schools, secondary schools and wharekura. Originally, our intention was to gather baseline data in the first year (Phase 1) and then collaboratively design an intervention that would improve both teaching and learning (Phase 2), which we followed for the primary and secondary school cases (see Moeed & Anderson, 2018). Instead of the two phases, we added an extra pre-phase for the wharekura case study as shown in Fig. 2.1.

Our model was for the teachers and students to learn science together in the first year (pre-phase), which involved a total of 10 two-hour sessions of science teaching and learning facilitated by the researchers. After the initial session, the teachers said they felt confident to start teaching Pūtaiao 2 hours a week using some of the pedagogies that had been modelled previously by the researchers for the teachers. In the second year of the project, participating teachers taught, and the researchers collected the data on both teaching and learning. We analysed and shared the data with the teachers who had time to reflect and read the relevant literature that we provided. Teachers teaching students to carry out science investigation followed this up. The process was a co-construction of learning to carry out a fair testing type of investigation (Phase 1).

What followed was teachers planning and teaching a variety of investigations including pattern-seeking, classifying, building models and providing opportunities for the students to explore (Phase 2).

The research design allowed both the researchers and participants the space for a bicultural framework. Although there are many examples of bicultural research in the mainstream system, this project was centred within te ao Māori by ensuring a Kura Kaupapa Māori philosophy was adhered to. The researchers were in the Kura learning from the teachers' ideas about Mātauranga and the teachers were learning science and approaches to Western science and investigation. One thought often expressed by Sue was that she felt the philosophy and practices of the kura were genuinely respected: 'You want for our kids what we want for them', more specifically, making it possible for the students to walk in both worlds—'theirs' and 'the other' (Rofe, Moeed, Anderson, & Bartholomew, 2016, p. 6).

2.6 Summary

This chapter presented the research design which is an interpretive case study into the teaching and learning of science investigation in a wharekura. The research design is responsive to science teaching and needs at the wharekura and is underpinned with a social constructivist theory of learning. During the planning and indeed throughout the research project, the principle of ako was practised. The researchers were learning about the wharekura, their philosophy, priorities and tikanga (way of doing/custom) and what it means to have the privilege to be involved in the science teaching and learning at the wharekura and indeed any kura that has a foundation of Kura Kaupapa Māori.

Question to consider
1. Critically evaluate the study design. What strengths and weaknesses do you see in the research design and methodology? If you were designing this research, what changes would you make and why?

Appendix 2.1: Observation Schedule

Observation Schedule: Date: **Day:**   **Time**

Purpose: What science investigation is taught and how?

   To record learning opportunities and student engagement during a science investigation lesson.

Classroom environment:
Physical aspects:
- Layout
- Lighting
- Room Temperature

Teacher welcomes students to class

Number of students in class

Demeanour

Lesson overview

Start of the lesson: topic/focus, style)
Uses an advance organiser
Learning intentions shared: Written/Verbal
Makes links to the previous lesson
Checks homework

Contextualisation

Looking for connecting with students; Students' world, Culture

Evidence of relationship with the class

Investigation:

Type: _____

Skill involved and developed:

Taught   Practised   Applied

How is the investigation introduced?

Teacher led   Demonstration   Posed as a problem   Student driven

Instructions

Written/verbal/on handout/From the textbook

Student involvement

How are students involved?
Individually/in groups [teacher or self-selected]
Decision about:
- What to investigate
- What equipment to choose
- What to measure
- How to record results
- How to analyse and summarise
- How to manage time? (Were they given time to think?)
- Critiquing their own design
- Critique others' design

How was the investigation concluded?
Reflection yes/No
What was the focus for the reflection?

Quality of engagement:

Enthusiastic: (Keenness/eager to start)	Most	Some	Few
Perseverance (Carrying it through)	Most	Some	Few
Attentive (*Behaving within the guidelines of class expectation*)	Most	Some	Few
On-task behaviour	Most	Some	Few

Researcher's view about what was learnt?

What was the focus Conceptual/Procedural/skill development?

What do you think student learnt?

Evidence:

References

Abrahams, I. (2009). Does practical work really motivate? A study of the affective value of practical work in secondary school science. *International Journal of Science Education, 31*(17), 2335–2353. https://doi.org/10.1080/09500690802342836.

Abrahams, I., & Millar, R. (2008). Does practical work really work? A study of the effectiveness of practical work as a teaching and learning method in school science. *International Journal of Science Education, 30*(14), 1945–1969. https://doi.org/10.1080/09500690701749305.

Abrahams, I., & Reiss, M. J. (2012). Practical work: Its effectiveness in primary and secondary schools in England. *Journal of Research in Science Teaching, 49*(8), 1035–1055. https://doi.org/10.1002/tea.21036.

Driver, R., Asoko, H., Leach, J., Scott, P., & Mortimer, E. (1994). Constructing scientific knowledge in the classroom. *Educational Researcher, 23*(7), 5–12. https://doi.org/10.3102/0013189x023007005.

Education Review Office. (2013). *Priorities for children's learning in early childhood services.* Wellington: Author.

Grbich, C. (2007). *Qualitative data analysis: An introduction.* London: Sage.

Merriam, S. B. (1998). *Qualitative research and case study applications in education: Revised and expanded from "Case study research in education.".* San Francisco, CA: Jossey-Bass.

Merriam, S. B. (2001). *Qualitative research and case study applications in education.* San Francisco: Jossey-Bass.

Millar, R. (2011). Reviewing the national curriculum for science: Opportunities and challenges. *Curriculum Journal, 22*(2), 167–185. https://doi.org/10.1080/09585176.2011.574907.

Millar, R. (2012). *Doing science (RLE Edu O): Images of science in science education.* Bristol: Francis Taylor.

Moeed, A., & Anderson, D. (2018). *Learning through school science investigation: Teachers putting research into practice.* Singapore: Springer.

Patton, M. Q. (2002). Two decades of developments in qualitative inquiry: A personal, experiential perspective. *Qualitative Social Work, 1*(3), 261–283. https://doi.org/10.1177/1473325002001003636.

Penetito, W. (2009). Place-based education: Catering for curriculum, culture and community. *New Zealand Annual Review of Education, 18*(2008), 5–29.

Rofe, C., Moeed, A., Anderson, D., & Bartholomew, R. (2016) Science in an indigenous school: Insight into teacher beliefs about science inquiry and their development as science teachers. *The Australian Journal of Indigenous Education, 45*(1), 91–99. https://doi.org/10.1017/jie.2015.32.

Sharples, P. (1994). Kura kaupapa Māori. In H. McQueen (Ed.), *Education is change* (pp. 11–21). Wellington, NZ: Bridget Williams Books.

Smith, G. H. (2000). Maori education: Revolution and transformative action. *Canadian Journal of Native Education, 24*(1), 57.

Solomon, J. (1987). Social influences on the construction of pupils' understanding of science. *Studies in Science Education, 14*(1), 63–82. https://doi.org/10.1080/03057268708559939.

Stake, R. E. (1995). *The art of case study.* Thousand Oaks: Sage.

Stake, R. E. (2005). Case studies. In N. K. Denzin & Y. S. Lincoln (Eds.), *Handbook of qualitative research* (pp. 435–454). Thousand Oaks, CA: Sage.

Taber, K. S. (2016). Constructivism in education: Interpretations and criticisms from science education. In E. Railean (Ed.), *Handbook of research on applied learning theory and design in modern education* (pp. 116–144). Hershey, PA: IGI Global.

Yin, R. K. (1994). *Case study research: Design and methods* (2nd ed.). Thousand Oaks: Sage.

Chapter 3
Teachers and Students Learning Through Science Investigation (Pre-phase)

3.1 Introduction

Knowledge required for teaching science and how this knowledge develops continues to be a focus of research in science education. In recent times, science curricula internationally aim for students to do science, learn science, understand how science works and develop scientifically literate citizens able to make informed decisions about socio-scientific issues that are likely to impact on people and societies. To deliver such a curriculum, teachers require subject matter knowledge, substantive knowledge (knowledge produced by science) and syntactic or epistemic knowledge (knowledge about science), the principles and process by which scientific knowledge is developed and accepted (Anderson, 2015). The knowledge about science is often referred to as the nature of science in school curricula. Access to the nature of science knowledge has been reported as being problematic for primary school teachers (Harlen, 2018). Teachers also need to know how to teach these two kinds of knowledge effectively, and they need pedagogical content knowledge (Anderson & Clark, 2012; Shulman, 1987). The participating teachers had considerable pedagogical content knowledge as they were experienced teachers, for example, this is evident in the way one participating teacher co-constructed the learning with her students, for example, see Fig. 4.1 showing the sequential process of planning, carrying out an investigation and reporting the findings. They also had the knowledge of their context, Mātauranga Māori and knowledge about their students.

Preparing quality science teachers is essential to ensuring students' success (Commission on Mathematics and Science Education, 2009; Darling-Hammond & Bransford, 2007; Schneider & Plasman, 2011). Researchers stress the need to think about how science teachers are prepared and some argue that often professional development experiences are disjointed and disconnected and do not always translate into enhanced teacher practice (Garet, Porter, Desimone, Birman, & Yoon, 2001; Schneider & Plasman, 2011). With these research findings in view, we designed a

© The Author(s), under exclusive licence to Springer Nature Singapore Pte Ltd. 2019
A. Moeed and C. Rofe, *Learning through School Science Investigation in an Indigenous School*, SpringerBriefs in Education, https://doi.org/10.1007/978-981-32-9611-4_3

professional development programme in the teachers' school, with their classes, where teachers, students and researchers collaborated and learnt together and from each other. Each new workshop was informed by critical reflection on the previous workshop.

This chapter reports on professional development for the participating teachers and the learning opportunities afforded to the students. In the pre-phase of the research, teachers and students were introduced to science investigations conducted by the researchers. Illustrative examples of these investigations are presented here, the focus being on teachers' learning and gaining confidence to teach, and for students learning how to investigate.

The two researchers planned and facilitated ten 2 h hands-on workshops for the two teachers on teaching and learning science in the junior school, where students were in Years 9 and 10 (aged approximately 13–14 years). The two teachers participated in the hands-on workshops alongside the students. As the students had not previously experienced formal science learning, engagement and enthusiasm for learning were high. The workshops were purposefully designed so that the teachers and students were learning the science ideas and at the same time the teachers experienced a variety of investigative approaches, details of which are presented in Table 3.1.

3.1.1   An Illustrative Example: Investigating the Speed of a Golf Ball

The first lesson started with watching a brief video of the 100 m final race at the London Olympics in 2012, which was won by Usain Bolt in 9.63 s. (https://www.youtube.com/watch?v=2O7K-8G2nwU). Students wanted to watch it again and it proved to be a useful hook for the ensuing discussion about the distance of 100 m covered in 9.64 s and the class working out the speed using the formula speed = distance/time.

This provided a useful lead into the lesson, where the learning intention was to work out the speed of a golf ball. Students were informed that they would go outside and use golf clubs and balls so that each group could find out the speed of their golf ball.

> Students shy or reluctant to answer questions until they were asked about the sports they played then offering suggestions about how we could find out the speed of the golf ball given to them. (Field notes)

The two teachers went around and talked with the students, who were more willing to talk with their teachers than the researchers. It was generally agreed that students needed to hit the ball, measure the distance it travelled and the time it took to travel that distance. Then the details were developed through questioning:

Table 3.1 Achievement objectives, content and planned learning outcomes of the workshops

Lesson	Achievement objectives	Science lesson (2 h)	Learning outcomes
1. School term 1	**Physical world**: *Physical inquiry and physics concepts*. Explore, describe and represent patterns and trends for everyday examples of physical phenomena such as movement, forces… **Nature of science**: *Investigating in science*	Investigating the speed of a golf ball	Students will be able to • Identify and describe the effect of forces on the motion of objects • Plan an investigation with guidance • Collect relevant data to measure speed • Use SI units of distance/time
2. School term 1	Build on prior experiences by working together to share and examine their own and others' knowledge Ask questions, find evidence, explore simple models and carry out appropriate investigations to develop simple explanations	Speed, mass, and acceleration	Students will be able to • Calculate the speed of an object with appropriate units
3. School term 1	**Material world**: *The structure of matter* Begin to develop an understanding of the particle nature of matter and use this to explain observed changes **Nature of science**: *Understanding about science* Appreciate that science is a way of explaining the world and that science knowledge changes over time	Structure of atom Elements, compounds and mixtures	Students will be able to • Explain atoms as small particles that make up all matter and identify subatomic particles • Make paper models to visualise what an atom may look like • Distinguish between an element and a compound, a pure substance and a mixture at the particle level • Apply simple separation techniques to separate a mixture of copper sulphate, sand, sawdust, and iron sand

(continued)

Table 3.1 (continued)

Lesson	Achievement objectives	Science lesson (2 h)	Learning outcomes
4. School term 2	**Material world:** *Properties and changes of matter* Group materials in different ways based on the observations and measurements of the characteristic chemical and physical properties of a range of different materials *Chemistry and society* Relate the observed characteristic chemical and physical properties of a process **Nature of science:** *Investigating in science* *Communicating in science*	Physical and chemical change	Students will be able to • Describe the three states of matter • Explain the process that leads to the change in state • Discuss the basic steps of the investigative process
		Acids and bases	Students will be able to: • Classify acids and bases through exploring their physical and chemical properties • Apply a simple framework to write up their investigation
5. School term 2	**Living world:** *Life processes* Recognise that there are life processes common to all living things and that these occur in different ways *Ecology*	Life processes	Students will be able to • Describe the life processes of all living things • Classify living and nonliving things • Use examples and discuss how classification is used in everyday life
6. School term 3	Explain how living things are suited to their particular habitat and how they respond to environmental changes, both natural and human induced **Nature of science:** *Understanding about science* Appreciate that science is a way of explaining the world and that science knowledge changes over time	Adaptations of snails	Students will be able to: • Explore snails • Ask questions and offer explanations to describe structures and behaviours of snails • Conduct guided inquiry into the needs of snails

(continued)

Table 3.1 (continued)

Lesson	Achievement objectives	Science lesson (2 h)	Learning outcomes
7. School term 3	**Living world:** *Life processes* Recognise that there are life processes common to all living things and that these occur in different ways **Nature of science:** *Investigating in science*	Exploring leaves and flowers	Students will be able to • Classify leaves based on their shape, colour and other physical features • Explain how classification helps to group similar things • Explore a flower to locate the different parts of a flower • Describe the function of different parts of a flower • Discuss similarities and differences between an assortment of flowers
8. School term 4	**Living world:** *Life processes* Describe the organisation of life at the cellular level *Environment* **Nature of science:** *Investigating in science* *Communicating in science*	Life at a cellular level	Students will be able to • Take the magic school bus to drill down from systems and organ level to the cell level • Describe the main organelles of a plant and animal cell • Make a paper model of either a plant or animal cell • Explain the similarities and differences between animal and plant cells
9.		Plant and animal cells	Students will be able to • View plant and animal cell slides • Use a binocular microscope • Prepare plant cell slides and view under the microscope

(continued)

Table 3.1 (continued)

Lesson	Achievement objectives	Science lesson (2 h)	Learning outcomes
10. School term 4	**Living world**: *Evolution* Explore patterns in the inheritance of genetically controlled characteristics Describe the basic processes by which genetic information is passed from one generation to the next Explain the importance of variation within a population **Nature of science**: *Participating and contributing* Develop an understanding of socio-scientific issues by gathering relevant scientific information in order to draw evidence-based conclusions	Genetics and variation	Students will be able to • Use physical characteristics to describe variations among the students in the class • Use a model to explore monohybrid crosses • Consider the reasons for these variations • Read Gene Seekers booklet to learn about a genetically inherited condition in the Māori population in New Zealand • Discuss ethical issues in carrying out medical research

Researcher: How many people do we need in each group?

Student: Three one to hit the ball, another in-charge of a stopwatch, and the third to walk the distance to work out how many metres the ball had travelled

The class of 12 students was organised into four groups of three students, who went outside to the sports field. Each group was given a golf club and ball to practise hitting the ball for 5 min. Each group had a stopwatch to measure the time; however, there was only one 200 m long measuring tape. Through teacher questioning, students decided that the 200 m tape was inadequate to measure the distance travelled by the ball and therefore the tape was laid out on the field and students counted the steps to cover 200 m in order to calculate the distance the ball travelled:

Teacher (Sue): Did everybody take the same number of steps?

Student: No.

Teacher: How can we make sure that our measurement of the steps is about the same?

Another student: No. Well … one person in each group could walk the 200 m and count their steps

The other teacher (Liz) suggested that it would be better if each student in the group was able to take turns with each of the roles. The teachers supported the students. The consensus was to have ten people, including the teacher, walk the distance and count the number of steps, then they could add them up and divide by 10 to get an average. It appeared that they had learnt to do averages in maths and students were able to work out the average number of steps that added up to 200 m. The students then took off to different parts of the field armed with a golf club, a golf ball, a stopwatch and a book to record the time and distance. Students enthusiastically went about collecting data and then returned to the class where each group worked out the speed of the golf ball. Then the data from all groups were collated on the board. Students copied this table into their books and then worked out the speed of the golf ball after which they were asked to work in groups to consider their findings. This led to a discussion about why they could not infer that the speed was the same for each ball. One student said 'Tom might have hit the ball very hard and Mary's ball did not go the same distance'. Another said 'the balls were all different … they may not be the same heavy … weight'.

At the end of each lesson, teachers and researchers reflected on the lesson, which was an opportunity for the teachers to ask any questions they had about the content or the teaching strategy. At the end of the above lesson, teachers and researchers reflected on the lesson and agreed that through questioning we could see whether students were beginning to develop some ideas about the investigative process. With support from the teachers and researchers, students had managed to plan and carry out an investigation and they tried to make sense of their data. However, we did not do any formal assessment, so this evidence of student learning is weak. Based on our observations and field notes, we can say that all students had been engaged, appeared to be enthusiastic and asked when they would be doing science again, the latter perhaps indicating that they were interested in learning science.

Sue's initial reaction was that back in the class there was a *lot of thinking* and she did not see it as hands-on. We agreed that

- student engagement was high, and it appeared that some students had played with a golf ball and club before;
- the students knew how to measure and record the data, although the table for recording the data had been provided;
- the students had done some exercises in maths and all groups were able to calculate the speed of their ball without any help other than a calculator with support, the students were beginning to think about their data in a thoughtful and critical manner.

The researchers received an email from the teachers indicating that in the following lesson they had taken the class outside to find out the speed of a basketball by following our model lesson. After two workshops, the teachers informed us that they had begun to teach two hours of Pūtaiao each week. We cannot say what was taught and learnt in those two hours as we did not observe this teaching. What we can say is that the teachers had gained the confidence to believe that they could teach Pūtaiao.

Sue initially talked about how she saw the students learning science that is reflected in the following quote:

> A kaumatua (elder) had gone eeling and he took them through the whole process of preparing the eel from the beginning to the end, which is science. What you are doing with our kids is science… knowledge is knowledge and who are we to say one is better than the other?

We think that in the beginning, Sue's view was that students learn science when the whole process is modelled to them, which is reflected in the above statement about students learning the whole process of catching and preparing the eel for kai (food). The statement also indicated that Sue wanted to let the researchers know the value of Māori knowledge along with science knowledge. She also understood that what we were doing with the students was also science.

What we found interesting was that when we provided the materials for the students to explore, they did not get into it and to start trying things out, which we found was common in English-medium schools. Perhaps this was because much of the teaching at the kura is teacher led. The students ask for guidance and direction from the teacher. It was possibly the notion of tuakana teina, the young ones learning from the older. We had to encourage them to get started and to get involved. We also wondered if the context of golf was an unfamiliar one. However, talking to the students it became clear that they have had the opportunity to play golf. So perhaps the reluctance was about deciding who will take the lead.

Another example was evident in the workshop on snails, where we brought snails of various sizes and trays on which students were to put snails to have a closer look using a magnifying glass. The students were reluctant, as was Liz, one of our teachers. No one wanted to touch the snails! We raised their interest by asking questions such as: Do snails have eyes? How do they find their way? Where do you think they live? What might they eat?

Suddenly, there was excitement on one table; what they observed was two snails preparing for mating! And questions about boy snails and girl snails were being asked. We learned that this was a good opportunity to suggest some things they might want to investigate. We talked about finding out if the snails like to move up or down a ramp. What might they prefer to eat? Did they like moist or dry conditions? One student decided to find out more about their eyes and tentacles. Next, they were picking the snails up and putting them on perspex, turning the perspex over to see how snails move.

By the end of the lesson, all students were exploring and making close observations about the snails, trying out ways to find out the *speediest* snail, explaining that they thought snails like to go up and not down, that they leave a trail behind. They also had new questions: Do they like lettuce or cabbage? Why don't we see them during the day? Do they like wet or dry places? One large lettuce and one cabbage leaf were put into the snail house for the students to check which one got eaten first. At the end of the lesson, they decided that they would like to keep the snail house for the rest of the week so that they could check the snail's preferred food and so the snail house was taken to Liz's classroom. In the next lesson, they reported that the snails preferred the lettuce because the lettuce leaf was eaten first and maybe they ate the cabbage leaf because they had no choice. This shows that these students have the cultural practice of looking after and caring for living things. Our students were investigating and so were the teachers.

The teachers' learning in this phase was related to developing their own understanding of the science ideas presented in each session. We found that the teachers asked questions when something was unclear. They joined in the practical activities and as the year progressed two things were evident. One that the teachers were using other resources to update their content knowledge and the other that the confidence they gained meant they were asking questions to check students understanding of the science ideas. More details about teacher learning are presented in Chap. 4.

Questions to consider
1. After reading this chapter identify three key messages about teaching and learning science investigation and discuss why you think these are important?
2. In your view, were the students motivated to learn, and what did they actually learn?

Appendix 3.1: Sample Analysis of One Lesson Taught by Researcher 1

See Tables 3.2, 3.3, 3.4, 3.5, and 3.6.

Table 3.2 The beginning of each lesson

Physical and chemical change		Students will be able to: • Describe the three states of matter • Explain the process that leads to the change in state (Physical change) • Discuss the basic steps of the investigative process
Acids and bases		• Classify acids and bases through exploring their chemical properties • Apply a simple framework to write up their investigation • Differentiate between physical and chemical change.
LESSON OVERVIEW		Lesson 1 Number of students 11 Time 2 h
Stated Purpose		Today we will learn about what matter is We will also learn some properties of matter
Teacher emphasis		Making observations Defining matter, states of matter Physical change; chemical change
Classroom environment		Positive, students follow instructions, attentive during instructions Teachers exploring alongside the students
Lesson overview	Advance organiser	Keywords used during the lesson written up on the board as they were introduced Matter, solid, liquid, gas Change in state, melting and freezing; solidification, evaporation and condensation
	Shared LI	Yes
	Linked to previous lesson	Yes
Contextualisation Evidence of connecting with students; students' world; Culture		Mutual respect
Evidence of relationship		Getting to know the students. Respected.

Table 3.3 Investigation related

Start of lesson	Researcher 1 Lesson 1
Type	Exploration, Sorting and classifying
How was it introduced?	Verbal instructions from the teacher Teacher led, followed by the opportunity to explore
Skills Taught/Practised/Applied [Be specific]	Noticing, making observations, comparing
Teacher led/Demo/	Guided exploration
Instructions (Verbal/written/Text)	Verbal

Table 3.4 Student involvement and engagement

Student involvement Decision about what to do:	Instructions given by the teacher
Investigation 1 (Diagnostic assessment) Exploration of ice	Exploration of several common materials (e.g. ice, wood, plastic, orange juice, candle, water, milk, pencil, etc.). Students asked to group these and explain their groupings); students grouped these as solid, liquid Students rub an ice cube between their hands and observe the ice changing into water Teacher asked the students to turn on the electric jug which has water in it. The students could observe water changing into gas. A student held a cold plate at the mouth of the jug and they could observe steam change into water
Chemical change	This led to the teacher explaining what each of these processes was called and students thinking about whether the matter needed to be heated or cooled for the change of state to take place Brief discussion about these changes being physical changes and why Students followed instructions to add a liquid to a powder in a test tube. The students smelt the liquid and decided it smelt like vinegar. Then they were asked to add it to the powder in a test tube and look closely Students observed fizzing, although they were not sure if the test tube smelt of vinegar or not. They repeated this experiment and came to similar conclusions

(continued)

Table 3.4 (continued)

Student involvement Decision about what to do:	Instructions given by the teacher
Acids and Alkalis	Students followed instructions to test acids HCL; HNO_3; H_2SO_4 using a universal indicator to observe the change in colour. Then repeated it with NaOH; $Ca(OH)_2$. In groups, they came up with a general statement about colour change The last activity was to use the universal indicator and to sort out whether four household chemicals were acidic or alkaline The lesson ended with students copying down a framework for writing up investigations which included aim, equipment, method, results and conclusion.
Choose equipment	
What to measure	
How to record results	
How to analyse/summarise	
Time given to think	Yes
Critique their own design	No
Critique others' design	No
Student engagement	Most engaged
On task behaviour	Attentive/interested/contributed
Quality of engagement	Most engaged

Table 3.5 Summary of the end of lesson reflection

Reflection:	A definition for the matter was copied by the students. With student help, the teacher drew a model showing how particles are arranged in a solid, liquid or gas. Students copied these in their books The teacher drew a triangle with solid, liquid, and gas at the three corners. Students were asked to select the words freezing, melting, evaporation, condensation. Then talked about dry ice, and how it changes from solid to gas and said this was called sublimation Through questioning and collaborative discussion, it was deduced that physical change was not permanent (water could be made into ice and ice melts to become water) Vocabulary development. It was obvious that many new words were presented, practised and used in this lesson Students could give examples of observations they had made Two students were able to explain in their own words what they thought the physical and chemical change was Students were able to write a general conclusion about the effect of acids and alkalis on the universal indicator Skilful questioning to elicit student answers

Table 3.6 Relevance and focus of teacher talk

Links to real life….	Using examples from life; everyday materials, e.g. water, ice, freezing, boiling Household examples such as dishwashing liquid, lemon juice, liquid cleaner, lemonade were used as unknown chemicals
Teacher talk focussed on…	Vocabulary development Substantive ideas around change of state, physical and chemical changes Making observation Giving students the opportunity to think The lesson gave students the opportunity to listen, talk, think and communicate Investigation teacher led

References

Anderson, D. (2015). The nature and influence of teacher beliefs and knowledge on the science teaching practice of three generalist New Zealand primary teachers. *Research in Science Education, 45*(3), 395–423. https://doi.org/10.1007/s11165-014-9428-8.

Anderson, D., & Clark, M. (2012). Development of syntactic subject matter knowledge and pedagogical content knowledge for science by a generalist elementary teacher. *Teachers and Teaching, 18*(3), 315–330. https://doi.org/10.1080/13540602.2012.629838.

Commission on Mathematics and Science Education (US). (2009). *Opportunity equation: Transforming mathematics and science education for citizenship and the global economy*. New York: Carnegie Corporation of New York.

Darling-Hammond, L., & Bransford, J. (Eds.). (2007). *Preparing teachers for a changing world: What teachers should learn and be able to do*. San Francisco: Wiley.

Garet, M. S., Porter, A. C., Desimone, L., Birman, B. F., & Yoon, K. S. (2001). What makes professional development effective? Results from a national sample of teachers. *American Educational Research Journal, 38*(4), 915–945. https://doi.org/10.3102/00028312038004915.

Harlen, W. (2018). *The teaching of science in primary schools*. London: David Fulton.

Schneider, R. M., & Plasman, K. (2011). Science teacher learning progressions: A review of science teachers' pedagogical content knowledge development. *Review of Educational Research, 81*(4), 530–565. https://doi.org/10.3102/0034654311423382.

Shulman, L. S. (1987). Knowledge and teaching: Foundations of the new reform. *Harvard Educational Review, 57*, 1–22.

Chapter 4
Teaching and Learning Science Investigation (Phase 1)

The previous chapter has reported how teachers and students learned to investigate together while the researchers took turns in facilitating the learning and making observations and interviewing students and teachers. In the next phase, the two teachers, Sue and Liz, taught science to their respective classes and the researchers interviewed the teachers, observed their teaching and conducted focus group interviews with their students. This chapter focuses on the following three aspects:

- Teacher beliefs about the nature of science investigation and why students should be taught to investigate.
- Teacher practice of teaching science investigation.
- Student learning through engagement in science investigation.

4.1 Teachers' Beliefs

As discussed in the previous chapter, teachers' beliefs are likely to impact on their practice and research evidence suggests that beliefs are a major influence in the development of knowledge for teaching (Anderson, 2015). It is argued that teachers' beliefs may influence the implementation of the curriculum and therefore what science students experience and learn is often dependent upon teachers' beliefs. Lowe and Appleton (2015) suggest that any professional development needs to give due consideration to teachers' beliefs. Congruence between espoused beliefs about science teaching and learning has been demonstrated previously among New Zealand primary teachers (Anderson, 2014, 2015). Strong connections between classroom practice and teachers' beliefs about science teaching and learning have also been noted in New Zealand, Australia and elsewhere (Fitzgerald, Dawson, & Hackling, 2013; Moeed & Anderson, 2018; Tam, 2015).

© The Author(s), under exclusive licence to Springer Nature Singapore Pte Ltd. 2019   43
A. Moeed and C. Rofe, *Learning through School Science Investigation in an Indigenous School*, SpringerBriefs in Education, https://doi.org/10.1007/978-981-32-9611-4_4

At the start of the project, it was unclear to us as to how Sue and Liz conceptualised science investigation and why they thought this should be taught or not. When talking about why science should be taught, Sue referred to parental expectations and they wanted to do well for the children. Although Sue saw that the learning of science would offer added opportunities to the students, she had a strong belief that Māori children should have a strong understanding of their culture, they need to develop a strong identity as Māori. She emphasised that students need to develop a Māori perspective. She articulated her beliefs as

> I suppose without putting Science down because I do not profess to know everything about science, but for a Māori person it is important to have a Māori perspective. For our students who come through it is ingrained in them, understanding Māori philosophy, so that when they have got a good understanding of that… (Sue, first interview)

Sue believed that in the previous 8 years of schooling her students did have an identity as Māori; they understood their culture and all that it entails for a Māori way of *being*. She held a strong belief that the kura would provide the tools so that the students could go out in the world and pursue any science. She said, our students have two perspectives, 'the perspective of the world they live in today and the perspective of the world that their *tipuna* (ancestors) carried through for them, the path is always wider for them'. Liz explained that her own experience of school science had not been a positive one: 'Just sat in a class and filled out a book, watch the teacher and then went to the next class … it was boring'.

Both teachers said they did not know much science but wanted to learn *for* their students, so that the students did not miss out. We understand that it was not that they did not know science; rather, they lacked confidence in teaching Western science. Teachers asked us to teach science in English, in a designated classroom. Even though one researcher was Māori, the teachers saw him as 'the science expert' rather than an expert in Māori knowledge and language. There was an appreciation from the teachers of the body of knowledge that science contains but they did not want the two knowledge systems (Western science and Māori knowledge) to 'mix'. After the first workshop, teachers began to teach two-hours of Pūtaiao each week in Te Reo Māori.

When the teachers were asked what they thought were the key elements of science investigation, Sue said

> I am not sure how to answer that. It is awesome that these children get the opportunity to have their hands-on, to experiment with it, pull it apart and put it back together, turn it upside down. Having you navigating them with a science world view, it allows them to get the right words that go with the right forms of experiments. Words like friction, forces, things like that don't mean a lot to them so when you have got someone supporting that while they are doing the experiments and we feed them the little hooks that they have in Te Reo Māori, then it makes the learning much richer. (Sue, first interview)

Sue's response showed the importance she placed in the students having the two worldviews, Māori and the other. Although Liz said she did not know what a science investigation was, the following quotation illustrates that she was able to explain the process of investigating quite well:

> Doing of the inquiry (investigation) is fine, but I think it is the ability to translate that and analyse it and relating what they saw back to the original question to form their conclusion. …I think science does this whole trial and error thing … it would be cool if schools could move into that and give kids free run to trial it. I want them to find out things for themselves rather than giving them the answers … Māori kids like to try things out, rather than being told, they get bored just listening. (First interview with Liz)

Liz appeared to understand the nature of science inquiry and explained it as an iterative process requiring evaluation of inquiry and being thoughtful about what changes may be required:

> Planning out what you are going to do, how you are going to do it, data gathering, data display, working out formulae, and then thinking what didn't work here, and what would have to be changed. … Last week we went out after we had done the golf experiment and we played basketball and figured out the speed of that ball… (Second interview with Liz)

She explained the purpose of science and what her approach to teaching science would be

> Real science is about going out there and looking for new things, I go diving, I would love to take these children snorkelling, so they can explore… (Second interview with Liz)

The teachers continued to gain confidence and became more active participants in the lessons. Classroom observations and recording show that they were asking questions more often and they engaged in the hands-on activities more and more. One example of Liz's was a lesson where students were exploring the snails:

> The first reaction of the teacher was negative and she did not want to touch the snails. Once the students started to participate in the activities she became really interested in what they were doing, asking questions and encouraging students to observe closely. (Classroom observation notes)

4.2 Teacher Practices

Mansour (2013) has highlighted the consistencies and inconsistencies between teachers' beliefs and their practices. To understand how Sue's and Liz's beliefs translated into practice, we observed how their beliefs were enacted in the classroom. We witnessed the complex interactions that occurred between the teachers and learners before, during and after investigations and how they influenced opportunities to learn science. We saw many instances of student–teacher interactions, where students were asking questions and the teachers were encouraging the students to use their electronic devices to access information and the teacher helping the students to make sense of that information. We made six observations in each teacher's class during Phase 1 of the research. We present the evidence from these observations next.

4.3 Teaching the Investigative Process

Here, we present the evidence of the teaching practice in Sue's and Liz's classes:

Evidence from Sue's class

The teacher planned a unit of work before the beginning of Phase 1 and shared it with the researchers. Sue's plan was to teach students how to carry out a fair testing type of investigation. This was done in the context of acid and carbonate reactions. She planned to teach the process of investigating through a collaborative process and found the frameworks in textbooks a useful starting point. Rather than writing up the various aspects of investigation on the board and explaining, she developed these with student input. For example, Sue began by explaining to the students that they were going to learn about planning their own science investigations. She put up a large sheet of paper and wrote 'Purpose of investigation in the middle'. Then she asked the students to use their computers and lookup science investigation. As the students shared the information, they had searched for they talked about what the purpose of their investigation could be. One student came up with 'what is the idea we are testing?' They talked about various aspects that could be investigated such as acid–base reactions, which they had done in the previous year. Another student offered 'we can investigate what happens when a metal reacts with acid' (an investigation which they had done in the previous year). Sue wrote these ideas on the sheet. With students' help, this was converted into a researchable question. A similar process was followed to develop the method and how the results would be recorded; any conclusions would have to be based on evidence from the data collected and various aspects that were to be considered when writing a discussion (see Fig. 4.1).

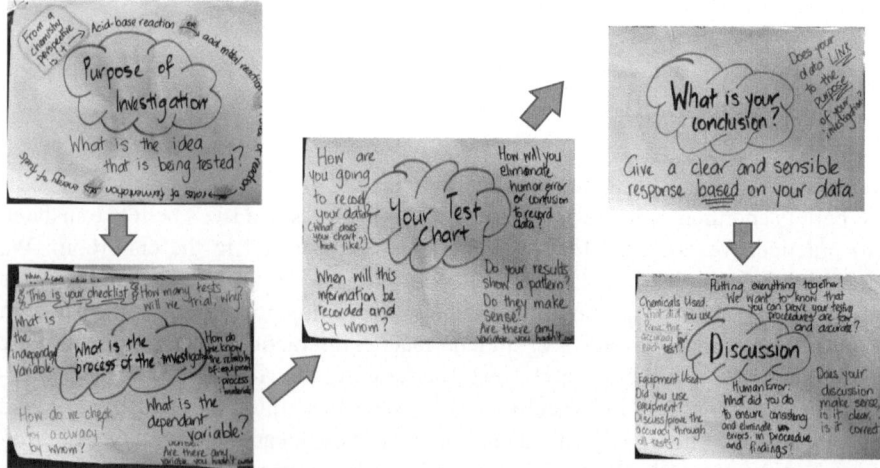

Fig. 4.1 Showing the sequential process of planning, carrying out an investigation and reporting the findings

The process was smooth because of Sue's skilful collaboration and her considerable teaching experience, her questioning, encouraging students to look for material on their computers and then talking it through to ensure student understanding came to the fore. Although the teaching was in English, the scaffolding used her own and students' knowledge of Te Reo Māori to clarify ideas and details. This use of familiar knowledge to construct new knowledge reflected her strong pedagogical approach (both Māori and non-Māori). Students were able to plan their investigation, develop a hypothesis, carry it out, make measurements, interpret and analyse data. All of these facets were achieved as a collective entity in the classroom where student individualism, which did exist, was not a focus for the cohort. Student success was 'measured' by all succeeding in classroom tasks. Most students were able to draw evidence-based conclusions from the data they collected. Many could explain a fair testing type of investigation, although they tended to view the investigation as a linear rather than an iterative process. However, although students were developing their understanding of the fair testing process and had experienced other approaches to the investigation such as exploration, classifying and sorting, pattern-seeking, and making and using models, the teachers and students generally did not consider these to be an investigation.

The teachers themselves were learning substantive science ideas alongside the students. Digital technology was a strong focus in the wharekura and teachers used it to access science knowledge with ease. They encouraged students to find out content using the many online resources available.

Both teachers had an added complexity to negotiate because their students needed to accommodate Western ways of making sense of their world that was inclusive of a Mātauranga Māori understanding of the world (Māori knowledge and Māori science). Both teachers appeared capable of creating an environment in which students felt their identity and views about their world and how it works were incorporated.

During the week, Sue contacted a researcher because she needed to 'get her head around' working out how to explain the collision theory to the students and to teach them how to write equations. She made time on the following Sunday to come and visit the researcher at home where an afternoon was spent learning how ions are formed and how to write word and chemical equations. We used formula card jigsaws to write formulae (see Fig. 4.2).

By the end of the session, Sue was writing up the formulae she would need to use with the students, and importantly how these cards could be used to teach her students to do the same. Her parting comment was, 'We were not taught like this at school, or I would have understood science'.

Later, Sue used her familiar approach to support students to 'look up collision theory' and they simplified and clarified it together as shown in Fig. 4.3.

Sue included the nature of science ideas while teaching science content. Together, the teacher and students unpacked and critically reflected on what scientists do when they investigate. Initially, Sue was more comfortable in teacher-led investigations. Her pedagogical approach to the investigation was 'being shown how to do something' (Rofe et al. 2016).

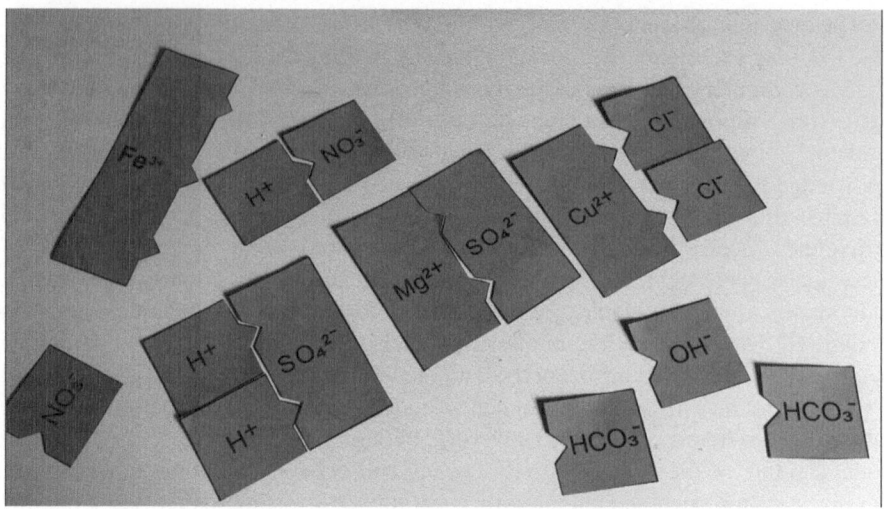

Fig. 4.2 Using jigsaw cards to write formulae

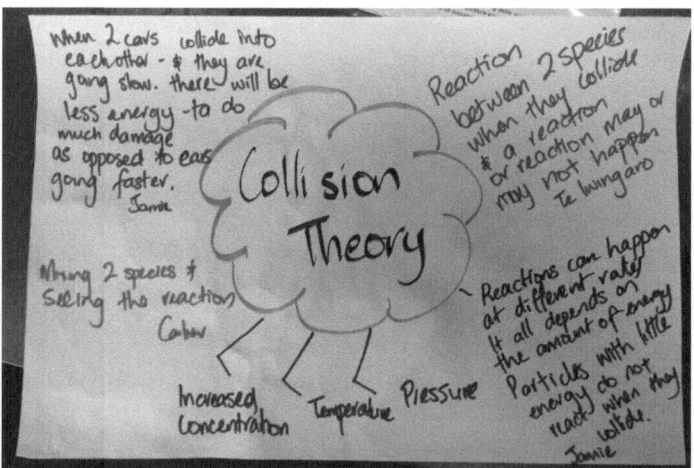

Fig. 4.3 Collaboratively working out the collision theory

Students continued to plan and carry out investigations, actively critiquing each other's plans and results, and were cognisant of the need for evidence to back their claims, asking 'How do you know?' 'What data did you collect?' and telling another group, 'You should repeat it to make sure you are right'. The critiquing process was supportive and there was much use of humour. In kura, students treat each other as whānau and are able to work together and laugh with and at each other.

Meanwhile, in Liz's class, the focus was on research investigations rather than practical science investigations. This was possibly because Liz was a social science teacher who was more familiar with inquiry learning. The following excerpt is about students selecting an animal of their choice and using their chrome books to find out specific information about the animal they had chosen. How she started the lesson to provide a focus for the students is illustrated in the excerpt:

Liz begins the lesson by writing on the board:

Te Ngahere (The Forest)
He aha tana kai? (What does your animal eat?)
Ko wai ka kai l a ia? (Who eats your animal?)
He aha te ahua o tana kainga? (What is the nature of your animal's home?)
He aha te taiao mai hei oranga mana? (What is the nature of where your animal lives?)

Next, Liz organised groups of three; there were some students in the class from a neighbouring class and they were included in the groups. She explained why those students were joining the group. Liz explained the questions in terms of kaitiakitanga (guardianship) of the Ngāhere (forest).

A student asked 'Are turtles native?' She responded 'Nga kararehe o te whenua' (Yes, animals of the land).

One student asked the researcher, What about kunekune (pig)? The researcher responded they are also animals of the land. Another student asked, What about penguins? This followed a discussion with the researcher about the pronunciation of 'hoiho' (yellow-eyed penguin and a hōiho horse).

Interestingly, the teacher was talking only in Te Reo Māori and the students were responding in English. Most of the time, students spoke to each other in English using Te Reo Māori from time to time. The students were encouraged to find out information for themselves and the teacher helped them to understand the information they were accessing. There was an awareness that although students could use a search engine to *access knowledge*, it was the teacher's task to help them understand it. Although Trinick and Dale (2015) argue that 'ako a kaka' or rote learning is an accepted methodology in Māori culture and suggest that this is done for the sake of precision and accuracy, the students were learning science with understanding.

Liz reminded the students about time expectations 'E rua miniti a te patai' (Two minutes before questions). The students had researched the following animals:

- Mokomoko (lizards),
- Kiwi,
- Toroa (albatross),
- Hoiho (penguin).

Liz used the large screen to display students' work on the chosen animals. The presentations were in Word format with pictures added in from the Internet (see Fig. 4.4). All words used were in Te Reo Māori. Each group presented their information to the class.

Fig. 4.4 Animals selected
by the students to investigate
and report on

The lesson was about investigating forest animals and looking at the producer—consumer relationships. All students used digital devices and appeared to have the requisite skills. The students were encouraged to communicate their findings to the rest of the class in Te Reo Māori. There was fruit in a bowl and the students were free to help themselves and eat if they were hungry while learning.

4.4   Summary

The teachers had gained confidence and used their experience and pedagogical skills to teach students how to investigate. In Sue's class, learning about the investigative process was through collaboratively accessing and unpacking the information collected whereas Liz scaffolded her students to carry out research and present their findings. Although practical investigations were common in Sue's class, the approach was mostly learning to carry out a fair testing type of investigation. The kura environment provided an ideal opportunity for students to be critical and creative in their investigative work. We noticed that the teachers were confident, they planned and prepared well, and were comfortable having researchers in the class and reflecting on the lesson. Our relationship with the teachers had developed to a point where they were no longer reluctant to ask for help and there was mutual trust, as Sue put it, 'You want for our tamariki what we want for them'.

It appeared, we were all together in the *learning waka*, paddling towards a common destination of students learning how to investigate in science. This imagery was used as a school-wide metaphor for the learning and journey of the kura.

Questions to consider

1. How would you include the two worldviews expressed by the teachers in this chapter? What relationships would support your understanding and validation of Mātauranga Māori?
2. What have you learnt about Māori pedagogical approaches used by these teachers?

References

Anderson, D. (2014). What makes it science? Primary teacher practices that support learning about science. *Set: Research Information for Teachers, 2014*(2), 9–17.

Anderson, D. (2015). The nature and influence of teacher beliefs and knowledge on the science teaching practice of three generalist New Zealand primary teachers. *Research in Science Education, 45*(3), 395–423. https://doi.org/10.1007/s11165-014-9428-8.

Fitzgerald, A., Dawson, V., & Hackling, M. (2013). Examining the beliefs and practices of four effective Australian primary science teachers. *Research in Science Education, 43*(3), 981–1003. https://doi.org/10.1007/s11165-012-9297-y.

Lowe, B., & Appleton, K. (2015). Surviving the implementation of a new science curriculum. *Research in Science Education, 45*(6), 841–866. https://doi.org/10.1007/s11165-014-9445-7.

Mansour, N. (2013). Consistencies and inconsistencies between science teachers' beliefs and practices. *International Journal of Science Education, 35*(7), 1230–1275. https://doi.org/10.1080/09500693.2012.743196.

Moeed, A., & Anderson, D. (2018). Science investigation in secondary school. *Learning through school science investigation* (pp. 71–91). Singapore: Springer.

Rofe, C., Moeed, A., Anderson, D., & Bartholomew, R. (2016) Science in an indigenous school: Insight into teacher beliefs about science inquiry and their development as science teachers. *The Australian Journal of Indigenous Education, 45*(1), 2015, 91–99. https://doi.org/10.1017/jie.2015.32.

Tam, A. C. F. (2015). The role of a professional learning community in teacher change: A perspective from beliefs and practices. *Teachers and Teaching, 21*(1), 22–43. doi.org/https://doi.org/10.1080/13540602.2014.928122

Trinick, R., & Dale, H. (2015). Head, heart, hand: Embodying Maori language through song. *Australian Journal of Music Education, 3*, 84.

Chapter 5
Enhancing Student Learning Through Science Investigation (Phase 2)

5.1 Research Informed Change in Teacher Practice

At the end of Phase 1, research findings were shared with teachers at a hui (meeting). We reported that participating teachers had successfully planned and efficaciously taught both science content and how to investigate. They had planned an investigation together and had asked the researchers for comments on their planning, seeking help if it was needed, and they gained confidence. As indicated in Chap. 4, participating teachers had considerable knowledge of their school, students and what it was to be a Māori teaching Māori students in Te Reo Māori. Both Sue and Liz were computer literate and had accessed relevant scientific information on the Internet from the Ministry of Education website and from other sources. We also reported that Sue had focussed on teaching a fair testing type of investigation, and there was evidence that her students could plan and carry out this type of investigation. Liz had involved her students in science investigations that were mostly research based and, as a result, her students did not experience as many practical investigations.

We reminded the teachers that through practical work and investigations students *learn to make connections* between the two domains of knowledge—the domain of real objects and observable things and the domain of ideas as illustrated in Fig. 5.1 (Millar, Tiberghien, & Le Maréchal, 2002).

At the hui, we provided teachers with scholarly readings including Abrahams and Millar's (2008) and Abrahams and Reiss' (2012) articles. These scholars report that although practical work and investigations are common in secondary schools, there is little research evidence that students learn from participating in practical work in general (science investigation is a subset of practical work). This comes as a surprise to most teachers including Sue and Liz. Based on their research, Abrahams and Millar claim that teachers do practical work because it is motivational, and students engage in it because they consider it to be a better option than writing. They also suggest a framework for analysing the *effectiveness* of practical work (see Fig. 5.2).

Practical work is effective at level 1 when students are able to *do* what the teacher intends them to do. However, the learning outcomes the teachers have in mind are

© The Author(s), under exclusive licence to Springer Nature Singapore Pte Ltd. 2019   53
A. Moeed and C. Rofe, *Learning through School Science Investigation in an Indigenous School*, SpringerBriefs in Education, https://doi.org/10.1007/978-981-32-9611-4_5

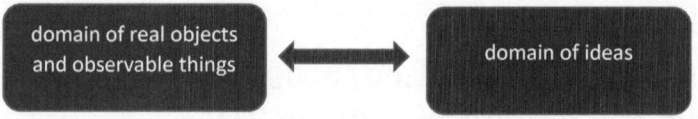

Fig. 5.1 Showing the role of practical work and investigation (Millar et al., 2002)

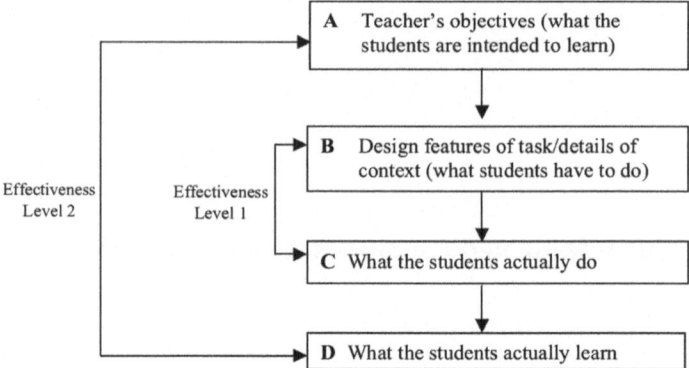

Fig. 5.2 Abrahams and Millar's (2008) framework for analysing practical work (p. 1947)

often not met at level 2; thus investigation is not effective. Abrahams and Millar suggest that practical work can be effective at level 2 if the teachers share the intended learning from an investigation with the students and include time for reflection at the end of the investigation. They advise teachers not to have too many intended outcomes for a single activity because it reduces the chance of anything being learnt. This framework and how it might be used was discussed with the teachers.

Our research project was collaborative. We did not impose an intervention but gave time for the teachers to look at the findings, read research articles, and decide what aspect of their teaching practice they would like to improve on. They suggested the following two changes they would like to make:

1. Providing an opportunity for students to experience a variety of approaches to science investigation as intended by the curriculum (for example, classifying and identifying, pattern-seeking, exploring, investigating models, fair testing, making things or developing systems).
2. Sharing the intended learning with students and making sure that this has taken place.

We present below the evidence of student experiences and effectiveness of investigation at levels 1 and 2 in both Sue's and Liz's classes.

5.2 An Investigation in Sue's Class

Sue chose to teach a sequence of lessons about microorganisms through a range of strategies including pictures, video clips and paper-based resources. She taught the content so that students could gain an understanding of microorganisms. She gave students the opportunity to grow microorganisms on agar plates and then view them using hand lenses. The focus here was for students to look at the microorganisms and identify them using patterns of growth (see lesson plan in Appendix 5.1).

Although the plan says that students would be pattern-seeking, it was more about students identifying and classifying microorganisms. It was also noticeable that there were achievement objectives both from the Living World and Nature of Science strands of the curriculum (Ministry of Education, 2007).

We observed a sequence of three lessons: first learning about microorganisms; second, learning about agar plates and how to inoculate the plates using cotton buds and samples taken from various places around the room; and third checking the agar plates to identify the microorganisms with teacher help and looking for patterns between each other's agar plates. Each lesson ended with focussing on what the students had *done*, and what they had *learnt*.

5.3 An Investigation in Liz's Class

In the selected lesson, the students' task was first to learn about some things that animals could do and how they were different from plants. The next activity required them to sort cards on which there were pictures of animals, plants or other objects. They were asked to sort the picture cards into animals, plants and neither. This created considerable discussion among the students working in each group. Finally, they had to give their explanations in writing. The lesson was structured around diagnostic, formative and summative assessment (Appendix 5.2).

> Students are able to differentiate between plants and animals but experienced difficulty with pictures of things that they had never known, for example, a picture of a wheat seed head. Student engagement is high and students are trying to group various animals into mammals and birds (Observation notes).

Interestingly, this was a hands-on activity and at the end, students were able to offer some evidence-based explanations.

Other points of interest were that the teacher had learning intentions for developing content knowledge and hands-on activities and there was time for the students to provide their reflections in writing. The planning also shows the extant science vocabulary that both the teachers and students had to learn. Learning vast amounts of science vocabulary has been considered a barrier to learning science for Māori students when they have to learn the Te Reo Māori words for the science terms (Stewart, 2007).

5.4 Analysis of Classroom Observation Data

Millar's (2010) framework was used for analysing, assessing and improving the effectiveness of practical science activities to analyse three observed lessons from each participating teacher's class (see Framework, Appendix 5.3). As shown in Table 5.1, teachers focussed on conceptual understanding and this was checked at the end of each lesson (two rows indicate more than two science ideas in the lesson), for example, students being able to define weight as a force and then explain the difference between mass and weight (see the key for the codes below the table). Similarly, a general understanding of science investigation was coded when there was evidence that a student had a grasp of the investigative process, they were able to write an investigable question, and showed knowledge of the need to gather evidence, process it and write a conclusion. Some students were able to offer explanations based on their results.

As these were small classes, keeping an accurate record was relatively easy. In some lessons, the reflective tasks were done using computers and the teacher could check students' responses. These were discussed at the post-lesson interview with the teachers. The students were looking for patterns and were able to identify similarities, differences, trend, and relationships with competence in Sue's class and in at least two of the three observed lessons in Liz's class.

Table 5.1 Student learning through investigation (from classroom observations)

Teachers	Sue				Liz	
Investigation number	1	2	3	4	5	6
Students can recall an observable feature of an object, or material, or event.	✓	✓	✓	✓	✓	✓
	✓		✓	✓	✓	
Students can recall a 'pattern' in observations (e.g., a similarity (s), difference (d), trend (t), relationship (r).)	✓	✓		✓		✓
	s, t	t, r	d, r	s, t		r
Students have a better understanding of a scientific idea (c), or explanation (e), or model (m), or theory (t).	✓	✓	✓	✓	✓	✓
	c, e	c, t	c, m	c, m	c, r	c, e
Students have a better *general understanding* of science investigation	✓	✓	✓		✓	✓
Students have a better understanding of some specific aspect of scientific enquiry	a, c, f	f	c, d	a, b, c, e, f, g	F	e, f
	a, b, c, f	a, b, d	c, b, d	a, b, g, i	d, e, i	g, h

Key:

a How to identify a good investigation

b How to plan a strategy for collecting data to address a question

c How to choose equipment for an investigation

d How to present data clearly

h How to critique evidence

e How to analyse data to reveal or display patterns

f How to draw and present conclusions based on evidence

g How to assess how confident you can be that a conclusion is correct

i How to communicate findings

Note Codes h and i have been added to the original Millar (2010) framework

5.5 Students' Experiences of Different Approaches to Investigation

Sue began Phase 2 with a fair testing type of investigation to ascertain that students recalled the process followed in Phase 1. All students were keen to respond to their teacher's questions. The investigation was related to the reaction of a metal carbonate (baking soda) with an acid (vinegar). Students were then given time to explore and make observations about what would happen when baking soda was added to vinegar. The following excerpt is from classroom observation.

Sue:	What did we decide vinegar was?
Iwa:	Acid (*tentative*).
Sue:	Do you all think it is acid?
Several students:	Yes, and baking soda is a base.
Sue:	Base?
Tane:	Yes, we can say it is an alkali because soda dissolves in water. Alkali is a base that is soluble in water

The students were asked to consider how they would find out what happens when different amounts of baking soda are added to vinegar. They each planned their own investigation but their plans, in fact, were quite similar. They were reminded to use their safety goggles and Sue checked each student's plan after which they could gather the equipment and carry out their investigation (Fig. 5.2a, b). One student decided to put the baking soda and vinegar in a plastic bag to time how long it took for the bag to pop (Figs. 5.3, 5.4, and 5.5).

In the following lesson, two students wanted to bake muffins using different amounts of baking soda. They had made their plan, found a recipe and were allowed to make muffins. They used four different colours of muffin paper cups to differentiate between each group of muffins, depending on how much baking soda they had and also included white cups as a control where they used the amount suggested in the recipe. They collected the data from photographs of the muffins rising, and they included a taste test. The students concluded that more baking soda makes muffins rise more but, when tasted, the muffins that had more than one teaspoon of baking

Fig. 5.3 Planning and carrying out an investigation

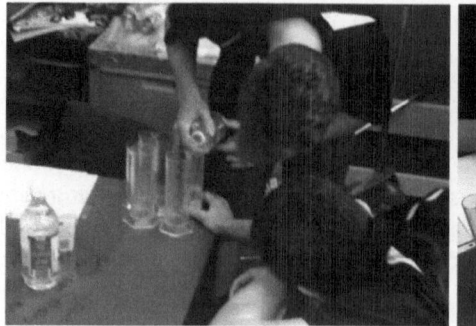

Fig. 5.4 Measuring vinegar and baking soda accurately

Fig. 5.5 Student timing the
bag to pop

soda tasted bitter. The teacher and the researchers who had the privilege of doing the
taste test agreed.

In the next lesson, Sue talked about factors which could affect the rate of chemical
reaction, reminding the students about the collision theory they had learned about.
The class collectively decided to investigate the effect of temperature on the rate of
chemical reactions. A table was co-constructed, and Sue checked that all students
knew what they were going to do (see Appendix 5.4).

5.6   Building a Model—Another Approach to Investigating

Sue also engaged her students in building a model as seen in Fig. 5.6. This observed
lesson started with a teacher-led discussion about frames. In humans, the skeleton is
the frame that supports the body. Crabs have an outer shell that provides support and

Fig. 5.6 Showing Hana's model of building a pātaka

protection from predators. Plants are built with cellulose, which provides a structure but some climbing plants do not have a strong stem and need other plants and stakes to support them.

The students were given paper and sellotape to build their structures. Māia first researched how pātaka kai are built. These are buildings on stilts, where traditionally Māori people stored food. She rolled each sheet of paper into a stick and made eight sticks to make two square frames but found that this was not strong enough so put one more stick in the middle to make it stronger. She had read that a pātaka has six poles, which was helpful. The approach was very much 'construct and try the stability'. The next step was to build the actual pātaka with a roof and walls and using this she was able to construct. This gave this student the ideal opportunity to make and test a model. Making models is an investigative approach promoted by the curriculum.

Liz involved her class in *Bush builders*, a project organised by the Wellington Zoo which involves students exploring the school grounds to find out what animals are living in their surroundings and to consider the kinds of environments that are needed for these animals. The students were encouraged to think about the food and habitat requirement of the animals they had found. They also talked about the environmental conditions that would be needed to attract particular animals to the school grounds. Students had the opportunity to explore, identify living things and to classify them. The investigations were mostly from the Living World strand of the curriculum (Ministry of Education, 2007). Students' experiences and prior knowledge were valued, and both teachers used Te Reo Māori for giving instructions. The students were engaged but a few needed reminders to get on with the task on hand.

Several observations were noted. Both teachers were confident with the science content material they were teaching; they had good pedagogical practices which meant they could focus on the science. For example, their planning and preparation were excellent and the lessons were well structured; this was evident by their recalling the previous lesson, sharing learning intentions, and having a reflection at the end of the lesson. The post-lesson interviews showed evidence-based reflection.

5.7   Summary

Both Sue and Liz provided opportunities for students to explore, plan and carry out fair testing types of investigation, to make observations, and to ensure students were focussed to look for patterns. There were fewer learning intentions for each practical task, and good practices such as wearing safety glasses, measuring accurately, having multiple trials and making evidence-based conclusions were encouraged. Both teachers planned well and had the requisite equipment for students to use. The teachers mostly spoke to the students in Te Reo Māori although the students responded in English. Key messages were, *what did you do? What did you learn?*

> **Questions to consider**
> 1. One role of school science investigation is to help students to make links between the domain of real objects and observable things and the domain of ideas. Do you agree/disagree with this statement? Use examples to support your view.
> 2. The New Zealand Curriculum requires students to experience a variety of approaches to science investigation. Why do you think it is important for students to have these experiences?

Appendix 5.1

Plan: Pattern Seeking With Agar - Bacteria - Friend or Foe!	
Achievement Objectives: Understand that scientists' investigations are informed by current scientific theories and aim to collect evidence that will be interpreted through processes of logical argument. Develop and carry out more complex investigations, including using models. Show an increasing awareness of the complexity of working scientifically, including recognition of multiple variables.	**Learning Outcomes:** • Study the basic life processes of bacteria by growing bacteria on agar plates. • Compare bacteria with viruses and/or fungi. • Pattern seeking. • Carry out experiments where the variables cannot be easily controlled. • Observe and record variables. • Identify patterns that result from these variables.
	Note: *Once a pattern has been observed this may lead to other investigations in an effort to try to explain why a particular pattern occurs, and may lead to a classifying and identifying system.* *Pattern seeking can also help us create models to explain observations.*
Lesson Sequence	Steady temperature. Identifying most suitable conditions for growing bacteria. See changes– in a time series measure the area of the growth.
New Vocabulary/Terms Shapes of Bacteria	

Researching

Researching involves gathering and analysing opinions or scientific findings in order to answer a question or to provide background information to help explain observed events.

Research can also show how scientists' ideas have changed over time as new evidence has been found.

Students need to practise each stage in the research process:

Stage 1: Focusing and planning
Questions relevant to the direction of the research are generated.

Stage 2: Sourcing information
Appropriate resources must be found. Using a range of different sources of information helps ensure the ideas are those commonly accepted.

Stage 3: Analysis
The information needs to be organised and then analysed to ensure that valid conclusions can be drawn.

Stage 4: Reporting
Finally, the research must be reported. This can be done in various ways—for example, a demonstration, a poster, a video or a report.

Appendix 5.2

Lesson Plan: Classification of Living Things	
Achievement Objectives	**Learning Outcomes**
Group plants, animals and other living things into science-based classifications. Investigate and understand relationships between structure and function in living organisms.	1. Classify living things into separate groups. 2. Learn new vocabulary and terms. 3. Classify living things based on science-based classifications. 4. Write your explanation to justify your grouping.

Lesson Sequence

Diagnostic Activity **What do you know?**

- Using the pictures, classify the different forms into groups that you think they belong to.
- Label them and explain why you put them into those groups.

Formative Activity **What do you want to learn?**

- Match new vocabulary/terms with their meaning.
- Use the pictures to re-classify living things.

Summative Activity **What I learnt.**

- Write your own explanation for the groups that you have put them into and justify why you have grouped them this way.

New Vocabulary/Terms

Background information

vertebrates	animals with an internal skeleton. Have a backbone.	invertebrates	do not have a backbone.
balance	an even distribution of weight in order to remain upright and steady.	cellulose	the chemical that is a main constituent in the walls of plant cells.
centre of gravity	the point around which the weight of an object or body is evenly distributed	exoskeleton	skeleton that covers the outside of an animal's body or that is in its skin.
endoskeleton	skeleton that is the structure on the inside of the body.	joint	a point at which parts of an artificial structure are connected. A structure in the animal body in which two parts of the skeleton are fitted together.
lignin	a compound in the walls of some plant cells which makes the plants strong and rigid. Lignin forms 25-30% of the wood in trees.	mobility	the ability to move or be moved.
spanning	extending across, e.g., a bridge spanning a river from one side to the other.	framework	a structure built by parts fitted or joined together.
foundation	the act of founding, creating, establishing.		

(Only a sample of pictures are included)

Appendix 5.3

A checklist for analysing and comparing up to 10 practical activities
  1 Learning objective(s) or intended learning outcome(s)

	Activity number	1	2	3	4	5	6	7	8	9	10
	1.1 Objective **(in general terms)**(Enter '1' for **main objective**; '2' if necessary for a **subsidiary objective**)										
A	By doing this activity students should develop their knowledge and understanding of the natural world.										
B	By doing this activity, students should learn how to use a piece of laboratory equipment or follow a standard practical procedure.										
C	By doing this activity, students should develop their understanding of the scientific approach to enquiry.										
A1	Students can recall an observable feature of an object, or material, or event.										
A2	Students can recall a 'pattern' in observations (e.g., a similarity, difference, trend, relationship).										
A3	Students have a better understanding of a scientific idea, or explanation, or model, idea, or theory.										
B1	Students can use a piece of equipment, or follow a practical procedure that they have not previously met.										
B2	Students are better at using a piece of equipment, or follow a practical procedure, that they have not previously met.										
C1	Students have a better *general understanding* of enquiry.										
C2	Students have a better understanding of some specific aspect of scientific enquiry.										

For C2, ↓ rather than simply ticking √ box, enter letters to indicate the specific aspects being taught as follows:

a How to identify a good investigation	e How to analyse data to reveal or display patterns
b How to plan a strategy for collecting data to address a question	f How to draw and present conclusions based on evidence
c How to choose equipment for an investigation	g How to assess how confident you can be that a conclusion is correct
d How to present data clearly	h How to critique evidence i.e. communicating (Additions to the original framework)

Appendix 5.4

Table: Purpose of the investigation
The purpose of the investigation is to determine whether a temperature change in the water/vinegar solution will change the rate of reaction when baking soda is added.

Equipment needed	Chemicals needed
Thermometer Hot water Beakers Measuring cylinders Stop watch Zip lock bags Paper towels	White vinegar Baking soda Hot water

Trials	Quantities			Temperature	Observations
	Baking soda	Vinegar	Water		
Test 1 Base test or control test	1½ Teaspoon	125ml	60ml	60 ml Warm water _____ °C	
Test 2				60 ml Warm water _____ °C	
Test 3				60 ml Warm water _____ °C	
Test 4				60 ml Warm water _____ °C	
Test 5				60 ml Warm water _____ °C	
Test 6				60 ml Warm water _____ °C	

References

Abrahams, I., & Millar, R. (2008). Does practical work really work? A study of the effectiveness of practical work as a teaching and learning method in school science. *International Journal of Science Education, 30*(14), 1945–1969. https://doi.org/10.1080/09500690701749305.

Abrahams, I., & Reiss, M. J. (2012). Practical work: Its effectiveness in primary and secondary schools in England. *Journal of Research in Science Teaching, 49*(8), 1035–1055. https://doi.org/10.1002/tea.21036.

Millar, R. (2010). *Analysing practical science activities to assess and improve their effectiveness.* Hatfield: Association for Science Education.

Millar, R., Tiberghien, A., & Le Maréchal, J. F. (2002). Varieties of labwork: A way of profiling labwork tasks. In D. Psillos & H. Niedderer (Eds.), *Teaching and learning in the science laboratory* (pp. 9–20). Dordrecht: Kluwer Academic.

Ministry of Education. (2007). *The New Zealand curriculum.* Wellington: Learning Media.

Stewart, G. M. (2007). *Kaupapa Māori science* (Doctoral diss.). University of Waikato, Hamilton, NZ.

Chapter 6
Student Learning Through Science Investigation

6.1 Introduction

The science learning area of the *New Zealand Curriculum and Pūtaiao i roto i te Marautanga o Aotearoa* (Ministry of Education, 2007a, b) requires students to learn about 'features of scientific knowledge and the processes by which it is developed' (p. 28), and to carry out science investigations using a variety of approaches. The overall aim of this project was to support teachers as they taught Pūtaiao/Science and engaged students in science investigation to find out what students learnt from it. Here, we report on student engagement and learning during the three phases of the research and since the end of the project. Student engagement is an indicator of motivation and motivation is an 'essential pre-requisite and co-requisite for learning' (Palmer, 2009, p. 147). Students need to be willing to get started and then be motivated to continue to want to learn.

The evidence shared here has been collected through classroom observations, post-lesson focus group interviews, samples of student work, results for the first cohort of students who completed Year 12 Biology and Chemistry, and a final survey and interview of this last group of Year 12 and 13 students.

6.2 Student Engagement and Learning Through Science Investigation in the Pre-phase

The analysis of class observation data showed that both attendance and engagement in the practical activities was good. In a class of 12 students, there was only one lesson out of the ten in this phase in which fewer than 10 students were present. Perhaps this was because there was novelty and variety in what they were doing, factors that have been reported as being motivational in the context of school science investigation (Moeed, 2016).

© The Author(s), under exclusive licence to Springer Nature Singapore Pte Ltd. 2019
A. Moeed and C. Rofe, *Learning through School Science Investigation in an Indigenous School*, SpringerBriefs in Education, https://doi.org/10.1007/978-981-32-9611-4_6

In the first of the 10 two-hour lessons, the students appeared to enjoy watching a brief video of the final 100 m race at the London Olympics in 2012 won by Usain Bolt of Jamaica in 9.63 s (https://www.youtube.com/watch?v=2O7K-8G2nwU). This was a very engaging 'hook' for the lesson. The students asked for the video to be replayed as they were intrigued by the distance covered by Bolt in less than 10 s and talked about the distance covered in each stride. Key points from the lesson observation were

> All students were engaged and after about 10 min, most were keen to get involved in the actual investigation. They particularly appear to like the idea of going outside and hitting a golf ball to find its speed……
>
> Students are answering the questions and once organised into groups, moved to different parts of the field to carry out their investigation…
>
> Most students (10) are able to use the data they have collected to calculate the speed of the ball, Tama says he learnt to do this in maths. Teachers provide calculators and are helping two students who needed support to get started. (Observation notes, first lesson)

In the next 2 h lesson on acids and bases, the students learnt the chemical properties and used indicators to classify household chemicals as acids, bases or neutral. All names used in the excerpts are pseudonyms and the researcher is R:

R: What did you learn today?

Aria: We learnt about acids and that kind of stuff, what acids do to litmus. Red litmus stays red when we put acid on. Blue turns red. *Actually* it is pinky, not red. Funny they call it red.

R: What else can you tell me about acids?

Māka: Lemons and that have acid. They taste sour. We used another acid today, something chlorine…Nah, chloric or something….

R: Hydrochloric?

Māka: Yes, that one.

R: What did you do with that?

Niko: We put it together with metals. Copper and aluminium did not work. Iron didn't work either.

R: What do you mean, they did not work?

Niko: Didn't do nothing. There was no reaction.

R: Did you get a reaction with anything else?

Māka: Yes, magnesium powder, *and* magnesium ribbon.

R: Tell me what do you mean by reaction?

Niko: There was gas coming off. The test tube got really hot with the magnesium powder.

Aria: It got hot with the ribbon too, but not that hot. It made a gas, we put a match in it and it made a noise. Funny.

R: Do you know what the gas was?

Maaka: No

Aria: It was hydrogen, very light, lighter than air

Initially, the students were a bit reluctant to respond to the researcher's questions, but they did encourage their peers to do so. It appears that at the end of the lessons, most focus group students had retained some of the information learnt during the lesson; this was evident as some students were able to answer questions put to them in the following lesson. As we did not use a quasi-experimental method of pre- and post-tests, there is little evidence of long-term retention of knowledge. The focus in this phase was on student engagement and there was evidence of high engagement in all 10 lessons.

6.3 Student Engagement and Learning Through Science Investigation in Phase 1

As reported earlier in Chap. 4, in this phase Sue and Liz taught their respective classes at the start of the following year. Sue had a class of 12 Year 10 students who were taught in the pre-phase, while Liz had 14 students in her Year 9 class. As reported in Chap. 4, Sect. 4.3, Sue started the year by teaching students the process of planning and carrying out an investigation. In contrast to the previous year when a focus, explore and reflect approach was used, Sue chose to first teach the investigative process. Her approach was co-constructive and students were encouraged to use search engines to look up what a science investigation was. They were supported to unpack the information and the teacher summarised it. One common theme in each observed lesson was

> Students, listen and do what the teacher asks them to do and through questioning Sue ascertains that the information is understood. (Observation notes)

The six observed lessons in Sue's class were on metals and genetics. The students were able to plan and carry out an investigation, for example, they planned their investigation to find out what happens as the amount of baking soda is increased in an acid–base reaction. The students were also able to collect and record their data and could explain that multiple trials are needed to have confidence in their findings. They used different methods of data collection; for example, Hana recorded the time taken from when she added the baking soda to when the fizzing stopped (See Table 6.1):

R: What did you find out?

Hana: That the fizzing time increases when there is more soda….but not that much difference between 8 and 9 then there was 7 and 10 (points to the table).

R: Why do you think that might be?

Hana   Don't know…maybe I did not time it properly.

Aria: Or there was too much soda … and not that much vinegar.

Table 6.1 Hana's recording of her experimental data		Baking	Vinegar	Fizzing time
Result table				
Test 1	½ Teaspoon	50 mL	3 s	
	1 Teaspoon	50 mL	5 s	
	1½ Teaspoon	50 mL	8 s	
	2 Teaspoon	50 mL	9 s	
Test 2	½ Teaspoon	50 mL	4 s	
	1 Teaspoon	50 mL	5 s	
	1½ Teaspoon	50 mL	7 s	
	2 Teaspoon	50 mL	10 s	
Test 3	½ Teaspoon	50 mL		
	1 Teaspoon	50 mL		
	1½ Teaspoon	50 mL		
	2 Teaspoon	50 mL		

R: Tell me more?

Hana: Whaea (teacher) said that for reaction the soda bits… particles … hit each other.

Aria: Co-collide or something?

R: Yes, collide, so you think there was less collision? Why?

Hana   I think there was too much soda sitting at the bottom or maybe we took the time more correctly.

R   So, what will you do next?

Aria   Repeat again … but we don't have time, maybe next time.

R: One last question, how will repeating help?

Aria: We could take an average, and that might be better

Liz's class worked individually on their research-based investigations. All students were working in the 'Google Classroom',[1] where teachers can create classes, set assignments, mark and send feedback and see all student work done, and just in one place. It is a useful programme that integrates with Google Docs and Google Drive; Google Classroom makes teaching more productive and meaningful by streamlining assignments, boosting collaboration and fostering communication.

This platform allows for a teacher to 'drop into' their virtual classroom to check progress. Liz used this tool to monitor students' on-task engagement and to provide feedback. If students could not understand what they had accessed through the internet, she would physically go to them and discuss what they were doing. The project ended with the students making their presentations. All students appeared to be computer literate, and digitally savvy and there was a real engagement around using devices.

---

[1] https://support.google.com/edu/classroom/answer/6020279?hl=en.

The following conversation is from the focus group interview conducted at the end of the presentations.

R: What new things did you learn from today?

Kiri: What kiwis are afraid of…

Matiu: That penguins are native to New Zealand.

Rongo: Umm, ohh, I learnt about different umm, types of lizards that live in New Zealand.

R: Okay, and what about you?

Ria: I learnt umm, different types of animals that eat mokomoko (lizards).

R: Okay, excellent. How did you learn these things today?

Kiri: Ohh, from Ria (looks at her friend) and from the internet and stuff.

R: From the internet?

Kiri: Just through browsing Google.

R: Yeah, sure. So, what about all the information, how do you know that information is correct?

Matiu: Because there's like heaps of pages that says the same thing…

R: From multiple sources, good. Where is a really good place for you to search for information?

Māia: The government sites, … is it?

R: Government sites, which ones do you find useful?

Kiri: Department of Conservation, Zoo, Zealandia…

R: What about offline? You're talking about all sorts of online.

Māia: and books…whanau…

6.4 Student Engagement and Learning Through Science Investigation in Phase 2

In Phase 2, students appeared more confident and engaged more readily, and they used scientific terms more correctly. Once the teacher put a question to them, they were able to organise themselves into groups and plan their investigations. For example, they investigated the effect of temperature on the rate of chemical reactions.

> Students are working in pairs to investigate how temperature may affect how fast baking powder reacts with vinegar. Interestingly, they are doing this by adding 60 mL of water at a range of temperatures. Wonder if they have thought about diluting the vinegar by adding water? (Observation notes)

It appeared they had considered the effect of dilution, because during the interview Māia said that they had added the same amount of water in each test tube, 'So they will all be watered down the same'.

While investigating metals reacting with hydrochloric acid, the students made the following observations:

- Magnesium reacted with acid; magnesium powder reacted faster than ribbon.
- There was fizzing when acid was added to some metals like magnesium.
- Iron filings stank! (There was some sulphur mixed in it).
- Iron fillings appeared to jump around in the test tube.
- When tested the gas produced was flammable.
- Aluminium did not react.

Based on their observations, they concluded that magnesium was the most reactive, followed by iron, and the aluminium did not react at all and the gas produced was hydrogen.

Students used formula cards to make sense of the chemical reaction. The picture on the right in Fig. 6.1 shows that this student was trying to put two hydrogen ions together to make hydrogen gas, which is a limitation of this model as it does not work for covalent bonding where electrons are shared.

While learning genetics, the students extracted cauliflower DNA following a protocol provided by the teacher. Later, they were asked to draw a picture of DNA and label it (Fig. 6.2). It shows how students were integrating their Māori knowledge with the science they were learning. Two interesting ideas came across, first that the family tree is indeed drawn as an image of a growing tree where there are leaves, a stem that represents the trunk, and roots; second, the rungs of the ladder not only had chromosomes, pattern, change and mutation but also ancestors and whakapapa. Perhaps this was a reflection of the strong connection of Māori people in general that these students have with their ancestors. As we had collected this information as a sample of the students' work, we did not have the opportunity to probe further.

Kaea (a special term for a leadership position). Inclusion of the word kaea on the top of the picture is perhaps the notion of that process leads to developing as a leader.

An example of student investigation can be seen in Kiri's report shown below:

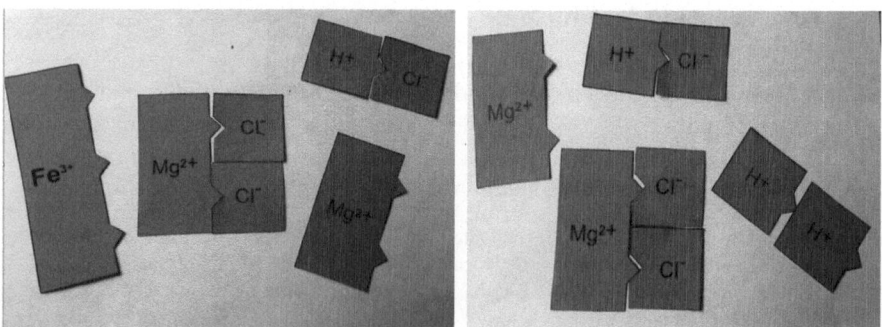

Fig. 6.1 Students using formula cards to make sense of the chemical reaction

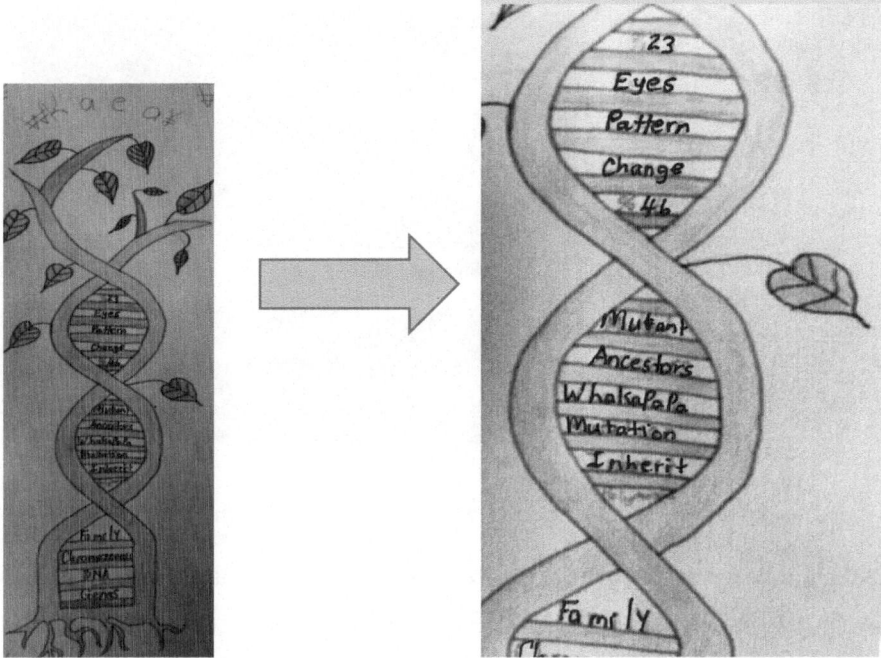

Fig. 6.2 Showing Rawiri's drawing of a DNA molecule

Te Iwingaro
What we did: Experiment #1

We put a spatula of Manganese di oxide in a test tube along with 5 ml of Hydrogen peroxide(with varying concentration with each test), we then blocked the top of the test tube containing the unknown gas that was create from the mixture. We then put a wood split in the test tube (that had been lit and blown out, with still glowing ember)

Test	Outcome
With low concentration of Hydrogen peroxide Test #1	The splint re-lit once: This indicated the gas created was a small amount of oxygen
With low concentration of Hydrogen peroxide Test #2	The splint re-lit four times: This indicated The gas created was, (while still only a small amount) a larger amount of oxygen than in test #1

Conclusion: The higher concentration of Hydrogen peroxide created a larger amount of oxygen, while the lower concentration created a smaller amount.

Another investigation one student did was to build a model of a building as discussed in Sect. 6.4 and shown in Fig. 6.3. Māia had researched how a pātaka is built, a structure familiar to her. Traditionally, food was kept in pātaka kai, a food storage building elevated from the ground as protection from rodents and other pests. She

Fig. 6.3 Showing Hana's
model of building a pātaka

decided to build a pātaka, a structure which was built on with three poles on two sides.

The following excerpt is from the focus group interview that followed:

R: What similarity do you see in a skyscraper and the structure you have built?

Māia: Both are built on a structure made of poles. Skyscrapers have an iron frame.

R: What gives these structures stability?

Māia: The number of poles and where they are put in. I think how deep the poles are put into the ground will also make them stronger and more stable.

R: How did you make your poles strong?

Māia   By rolling the paper tightly, I started by rolling around a pencil. Then used the sellotape to hold it together.

R: Did you do this by trying it out?

Māia: No, I learnt that at the BP technology challenge three years ago. There we made the poles with newspaper.

R: What other ideas that you have learnt help you in learning science?

Māia: I use harakeke (Flax) to make tukutuku (decorative wall panels) at the marae. Harakeke is stronger than paper. I like to make kono which I learnt to make in Taranaki; it can hold heaps of things. Harakeke is also used to hold the poles together, it is strong

When given the opportunity to investigate, Māia drew upon her Māori knowledge which gave a meaningful context to the science she was learning.

Liz's class had the opportunity to explore and learn about the plants and animals on their school grounds. For example, they learnt about the need to conserve the habitat and how to attract animals to their school grounds. They talked about their understanding of extinction.

R: What do you think the lesson was about?

Matiu: How to stop animals from extincting, how you save them.

Maia: We had to find out things about our animals.

Where they live? What does the animal eat? What eats the animal? What are the predators?

Kiri: To save the animals from extinction we have to make sure they have the right food and get rid of their predators.

Rua: I learnt about tuatara, their habitat, what they eat and why their numbers have decreased

Students had selected kiwi and kororā (penguins) as native animals that could be attracted by the school for the Bush Builders project. They understood that kororā needed to be close to the sea but that was not a possibility. They also knew that to have kiwi, they needed a predator-free habitat. The focus in Liz's class was mostly on the environment and sustainability. However, the students were learning Pūtaiao/Science and had an opportunity to explore, observe and offer evidence-based explanations.

6.5   Student Survey Results

In the pre-phase, the students had responded to a questionnaire which was repeated at the end of Phase 2. The survey was brief as we did not want the students to start their Pūtaiao/Science learning responding to a long questionnaire. As can be seen in Fig. 6.3, in both phases they saw having fun as the purpose of doing science investigation. The main difference between their responses in the two phases was that at the start of the project they had a naïve understanding of what the purpose of the investigation was but by the end they appeared to have a more nuanced understanding.

6.6   Student Engagement and Learning Beyond the Project

The first group of students taught by their teachers (Phases 1 & 2) sat the National Certificate of Educational Achievement in 2017. We sought consent from the students to access the results of those students who continued with Pūtaiao/Science and who did Chemistry and Biology in Year 12. Relevant to this research are the results presented in Tables 6.2, 6.3 and Fig. 6.4. (Note, Māori numbers are used as pseudonyms for student names). From these results, it is evident that students were able to carry out an investigation both in a biology and chemistry context and receive four credits in each case. This aligns with the curriculum requirement that students will be able to carry out more complex investigations with supervision (Level 7) and with direction (Level 6). We have also included their results from other achievement standards as evidence of these students performing well in both internal and external assessments. Although the Ministry of Education report for 2017 indicates that Māori students continue to underachieve, this has not been the experience of students who

Table 6.2 National certificate in educational achievement Year 12 biology results

Carry out a practical investigation in a biology context, with supervision

Candidate	Results	Credits	Internal assessment
1 Tahi	Excellence	4	
2 Rua	Achieved	4	
3 Toru	Achieved	4	
4 Wha	Merit	4	71%
5 Rima	Not Achieved		
6 Ono	Not Achieved		
7 Whitu	Achieved	4	

Demonstrate understanding of adaptation of plants or animals to their way of life

Candidate	Results	Credits	Internal assessment
1 Tahi	Achieved	3	
2 Rua	Achieved	3	
3 Toru	Achieved	3	
4 Wha	Merit	3	100%
5 Rima	Achieved	3	
6 Ono	Merit	3	
7 Whitu	Achieved	3	

Demonstrate understanding of life processes at the cellular level

Candidate	Results	Credits	External assessment
1 Tahi	Merit	4	
2 Rua	Merit	4	
3 Toru	Not achieved		
4 Wha	Achieved	4	85%
5 Rima	Did not appear for assessment		
6 Ono	Achieved	4	
7 Whitu	Achieved	4	

participated in this research as evident in the figures presented below (Tables 6.2, and 6.3).

> While outcomes for Māori improved in some areas, there continue to be persistent achievement and engagement gaps between Māori and non-Māori. Māori school students continue to achieve behind the nonMāori population for NCEA levels 1, 2 and 3 and University Entrance. (p.38[2])

It is evident that students who chose to continue with science at the research school were performing better than other students nationally.

---

[2]https://www.educationcounts.govt.nz/__data/assets/pdf_file/0005/188078/NZ-Schools-2017.pdf.

Table 6.3 National certificate in educational achievement Year 12 chemistry results

Carry out a practical investigation chemistry investigation with direction

Candidate	Results	Credits	Internal assessment
1. Tahi	Merit	4	
2. Rua	Merit	4	
3. Toru	Excellence	4	
4. Wha	Achieved	4	100%
5. Rima	Excellence	4	
6. Ono	Merit	4	
7. Whitu	Achieved	4	
8. Waru	Excellence	4	

Demonstrate an understanding of the properties of selected organic compounds (Level 2)

Candidate	Results	Credits	External assessment
1. Tahi	Achieved	4	
2. Rua	Achieved	4	
3. Toru	Achieved	4	
4. Wha	Achieved	4	100%
5. Rima	Achieved	4	
6. Ono	Achieved	4	

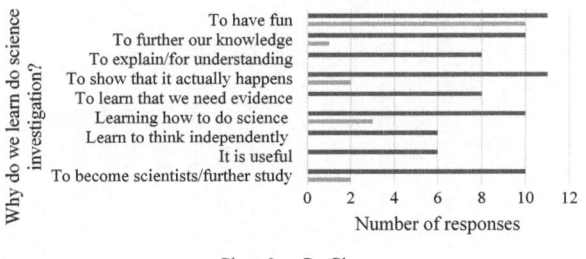

Fig. 6.4 Students' views about the purposes of science investigation

6.7 Since the Research Project

We followed this group of students and the science they were learning and the investigations they remember doing in the previous year. It gave us an opportunity of finding out why they had continued in science and what connectedness they saw between science and Mātauranga Māori. After gaining ethics approval, we invited the students to respond to a survey. The findings of the survey are presented in Table 6.4. Eleven

Table 6.4 Year 12 students' responses to survey questions

Questions	Number of Respondents	Examples	Teacher	Peers	Whānau	Online Programme like Education Perfect	Research	Homework
What science ideas do you remember from year 9 and 10?	20	Looking at cells under microscopes (×6) pH of acids and bases (×7) Water is neutral, pH 7 (×1) Any two snails can mate (×1) Friction slows things down (×1) Speed/time graphs (×4)						
What aspects of science work did you enjoy?	33	Learning new things (×8) Practical/experiments, investigations (×10) Making up circuits (×4) Using magnets (×4) Learning about native animals (×6) 'I enjoyed the idea of science and discovery; learning and understanding was the best' (Tāne) Calculations (×2) Working with mates (×4)						
What were some of the things that you did not enjoy?	6	Repeating things (×3) 'I am very particular when it comes to what I find interesting, so when I found something uninteresting it became bothersome' (Tāne) Doing calculations (×2)						
What do you think helps you to learn?	11	Practical investigations Experiments	9	8	2	11	8	0
Why did you decide to study Biology/Physics/Chemistry?	11	Physics Enjoyment (×1) Career options (×4)			Chemistry Enjoyment (×8) Career options (×2)		Biology Enjoyment (×6) Career options (×4)	

students who had continued with science responded to the survey (Table 6.4).

Student survey responses in Year 13

Tāne, who was studying Physics and Calculus in Year 13, knew what he wanted to do when he left school: 'I have had a long desire to be some type of engineer and learning sciences was one of the first steps to achieving that'. When asked about the connection the students saw between Māori knowledge and science, most students saw some connection between the knowledge of plants, their medicinal properties, using stars to navigate, and planting seeds according to the time of the year. Tāne's response was most articulate and comprehensive:

> Depending on what 'Māori knowledge we are talking about. If we talk about medicinal practices and care of one's body, then sure you could find some connection somewhere, but to talk about general building blocks of Māori knowledge and science knowledge, on the one side science bases itself on trial and error, constant and non-stop critique, and unwavering proof. Whereas on the other side Māori knowledge is based on tradition, faith, and the understanding of feelings rather than facts. I couldn't give a connection, except that I am learning both.

6.8   Summary

Student engagement was high in almost all lessons we observed. Evidence suggests that these students were learning science ideas as well as the science capabilities required to investigate. The classes were small and provided the opportunity for researchers to make closer observations and have more confidence in the findings in relation to student engagement and learning. In the pre-phase, our focus was more on student engagement that is reflected in the reported findings. Phases 1 and 2 showed that teachers were teaching science with confidence and students were learning.

Although a longitudinal study was not our planned intention, it has been useful to follow the first groups of students taught in Phases 1 and 2 of the research since the research project finished. Those students who wanted to continue with science did so and continued to learn.

Questions to consider
1. Discuss how the learning opportunities provided helped the students to draw upon their Mātauranga Māori to construct their Pūtaiao/science understandings?
2. What approaches to science investigation did the students experience, and what did they learn from them?

References

Ministry of Education. (2007a). *Pūtaiao i roto i te Marautanga o Aotearoa.* Wellington: Learning Media.

Ministry of Education. (2007b). *The New Zealand curriculum.* Wellington: Learning Media.

Ministry of Education. (2017). Ngā Kura o Aotearoa New Zealand Schools. Wellington. Retrieved from https://www.educationcounts.govt.nz/__data/assets/pdf_file/0005/188078/NZ-Schools-2017.pdf.

Moeed, A. (2016). Novelty, variety, relevance, challenge and assessment: How science investigations influence the motivation of year 11 students in New Zealand. *School Science Review, 97*(361), 75–81.

Palmer, D. H. (2009). Student interest generated during an inquiry skills lesson. *Journal of Research in Science Teaching: The Official Journal of the National Association for Research in Science Teaching, 46*(2), 147–165. https://doi.org/10.1002/tea.20263.

Chapter 7
Whakakapi (Bringing Together) Discussion

7.1 Introduction

There were three aspects to this research project and the findings have been reported in the previous chapters for the pre-phase, Phase 1 and Phase 2. The pre-phase had a professional development focus with students and teachers learning to investigate in science. One researcher did professional development and the other collected evidence of teacher and student learning. In Phase 1, teachers taught science and science investigation, and the researchers gathered evidence of their teaching it and of student learning. In Phase 2, the teachers focussed on improving their teaching of investigation and the researchers collected further evidence of teaching and learning to investigate. From our analysis of data, five themes emerged which are discussed in this chapter and are informed by the relevant research literature. We also present a model of professional development from our research.

7.2 The Emerging Themes

The five emerging themes are listed below and are discussed in the light of relevant literature:

1. *Teacher beliefs, agency and confidence;*
2. *Teaching and enhancing teaching of science investigation;*
3. *Student engagement and learning through science investigation;*
4. *Creating a place for Pūtaiao/science learning at the wharekura;*
5. *Ako reciprocity of researcher learning through the project.*

© The Author(s), under exclusive licence to Springer Nature Singapore Pte Ltd. 2019
A. Moeed and C. Rofe, *Learning through School Science Investigation in an Indigenous School*, SpringerBriefs in Education, https://doi.org/10.1007/978-981-32-9611-4_7

7.2.1 Teacher Beliefs, Agency and Confidence

The teachers believed that Pūtaiao/science should be taught and learnt by their students. However, both Sue and Liz were very clear that Mātauranga Māori and students developing their identity as Māori took primacy. They both knew that participating students had in their previous 8 years of schooling been taught in ways that supported them to develop their Māori identity, and most were fluent in Te Reo Māori. The kura (primary school) whānau wanted their students to learn science and this was the reason these two teachers were prepared to learn science and how to investigate. It was perhaps their own experience of school science, which they both said had not been a positive one, was the reason they lacked the confidence to teach science. They also said that they did not know any science, especially physics and chemistry. They gained confidence after participating in just two sessions in the pre-phase and began teaching 2 h of Pūtaiao each week (Rofe, Moeed, Anderson, & Bartholomew, 2016). Before the end of the year, Sue was motivated to do an online course focussing on the Nature of Science, which was offered by Victoria University of Wellington, to update her knowledge.

It was almost as though the teachers needed to be encouraged and offered timely support for them to learn science content and processes. Learning alongside their students and extending their own learning became common practice.

The first investigation they did with students by themselves was to find out the speed of a basketball, which was based on what we had modelled through our first session; a noticeable change was that Sue and Liz did not just repeat the ten sessions we had covered. In both Phases 1 and 2, they built on the content that students had already learnt and extended the students' learning. Sue had understood the purpose and process of science investigation and used her own approach to co-construct students' knowledge of science investigation with them. Liz used her approach to social inquiry to teach investigation. Both teachers ensured that students continued to learn Te Reo Māori by continuing to use it as the medium of instruction. Liz used science topics, which included environmental education, biodiversity, and sustainability, that were closer to her own experiences.

The teachers had the agency to choose the topics that were relevant and that fitted well within the framework of learning at the kura, and they upskilled themselves to teach the content in a confident manner. Although in Phase 1 Sue mostly focussed on a fair testing type of investigation, once the findings from Phase 1 were shared, both teachers created opportunities for students to experience a variety of approaches to investigation as recommended by Goldsworthy, Watson, and Wood-Robinson (1998) and the *New Zealand Curriculum* (Ministry of Education, 2007). Sue, with more experience and being given the task of developing a programme for science for Years 1–10 at the kura, put considerable time and effort into developing her own ideas about science investigation. Both teachers initially believed that science investigation was about hands-on engagement for motivational purposes but very quickly understood that *minds-on* was just as important as *hands-on*. This was obvious in Phase 1 when they shared the intended learning and checked for understanding at the end of each lesson. Initially, they saw the investigative process as linear and sequential; however, this changed in the second year when, for example, Sue encouraged students to revisit their plan and make changes based on what they had found out.

The teachers gained confidence, could access support in a timely manner, and had the agency to use the pedagogical approaches they knew worked in their context. They believed that students should learn science and took the time to gain the requisite content knowledge and skills so that their students had the opportunity to learn Pūtaiao/science. This model of learning together and accessing the support required for professional development worked well for this wharekura.

Traditional models of professional development are widely criticised as being ineffective due to insufficient time and content necessary for increasing teachers' knowledge and nurturing meaningful changes in their teaching practice (Abu-Tineh & Sadiq, 2018). An advantage of our model was long-term engagement, and it was tailored to the specific needs of the participants over an extended period. Consequently, the teachers gained confidence and set lessons in *their context*. The students were sometimes directed, and at other times selected their own investigations, planned, thought about what data they needed to collect, made their choice about how to proceed and interpreted their data.

Loughran and Berry (2011) contend that too often science teaching is in the form of delivering facts and information, and assert that good science teaching ought to support students to become responsible and active learners. Sue's statement, 'I don't tell them, just give them the space to think and work it out for themselves' is indicative of her pedagogical approach, which was encouraging her students to become responsible and active learners. Similarly, Liz had gained confidence and she too gave students the responsibility to conduct their investigations and monitored their learning.

7.2.2 Teaching and Enhancing Teaching of Science Investigation

In Phase 1, both teachers taught students how to investigate, the difference being a practical focus in Sue's class and a research focus in Liz's class. The researchers shared the findings from Phase 1 and the teachers had the choice of deciding what changes they wanted to make to their practice. Following Millar, Tiberghien and Le Maréchal's (2002) advice, they saw the purpose of the investigation as helping students to make links between the two domains of objects and observables and the domain of ideas (see Fig. 5.2). They appeared to be considering, *what science idea do I want the students to learn* and what would be the *best way for them to understand* it?

Having had time to read and think about the effectiveness of the practical work model proposed by Abrahams and Millar (2008), they were aware that the effectiveness of an investigation is dependent on what students learn from it (see Fig. 5.2).

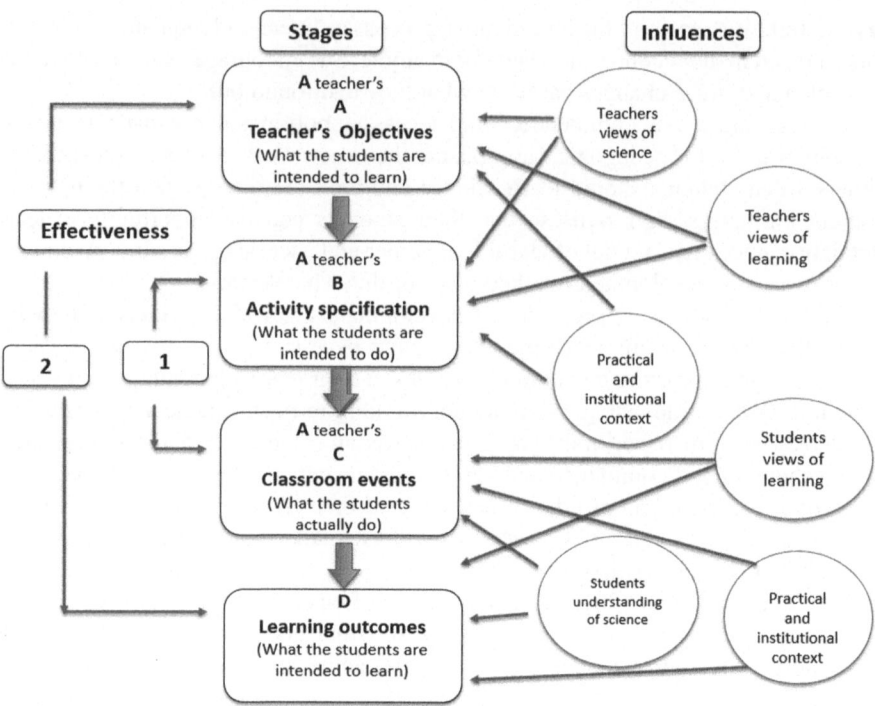

Fig. 7.1 Stages in developing, carrying out and evaluating a practical activity and the influences on these (Millar, 2010).

A common teacher practice in both classes was to share the intended learning with students at the start of the lesson and to have time at the end of the lesson for reflection to ascertain that the intended ideas and processes had indeed been learnt. In our view, their practice became a lot closer to that described by Millar (2010). In Fig. 7.1, Millar has expanded Fig. 5.2. The teachers were cognisant of the influences of both their own and their students' views about science and learning on the learning outcomes of the investigation. Both teachers had a deep understanding of their context and had the experience to know how their Māori students learnt. They gave their students time and space to think, discuss, plan, and carry out their investigations.

7.2.3 Student Engagement and Learning Through Science Investigation

In small classes, there is less opportunity for students not engaging and not being noticed. We did not see many instances where students were not participating. As the project progressed, students were willing to ask questions when they did not understand a science idea or were unsure about what to do. Millar (2010) asserts that student learning is affected by their views about their own learning, whether it is seen as constructing meaning from experience or by discovering by observing and measuring, or by confirming the ideas and insights given by the teacher. The analysis of our observations showed that students were learning science ideas and also learning the process of a fair testing approach to investigation. They knew that in such investigations variables needed to be controlled, that conclusions needed to be evidence based, and repetition or several trials made the results more reliable. Students used their senses to make observations, and were able to answer questions when asked *how do you know?* Teachers' practice of checking student work, as Liz did in Google Classroom and at the end of investigation reflections, provided plenty of evidence of student learning, which was regularly confirmed in the focus group interviews.

Although we do not have pre- and post-test marks, we have some evidence from the last survey we conducted of students' long-term retention of what they had learnt during the project. Even though there is little evidence of students learning from practical work internationally (Abrahams & Millar, 2008; Hodson, 1990), in our case participating students were learning through science investigation. These findings were similar to a case study research reported in New Zealand, which shows that when teachers are clear about the intended learning from an investigation and share this with their students and check that intended learning has been achieved, students do learn through science investigation (Moeed & Anderson, 2018). Murray (2007) analysed the achievement of Māori students in immersion and bilingual schools and reported that nearly 50% of all students who graduate from wharekura with National Certificate of Educational Achievement Levels 1, 2 or 3 do not get any credits at any level in Pūtaiao/science. The students at our wharekura were learning science at National Certificate of Educational Achievement Levels 1, 2 and 3 and gaining credits not only in the internal assessment of investigation but also in externally assessed standards. After the project, all students achieved in the internal assessment of science investigation for the National Certificate of Educational Achievement in Levels 1, 2 and 3 as reported in Chap. 6.

Osborne (2015) argues that 'Developing an understanding of an idea requires talking about it, writing about it, reading about it, and representing/drawing or visualizing it' along with engaging in practical work (p. 18). Students in this research were given opportunities to read, write, talk and visualise as they practised investigation. Social constructivist approach to learning from each other, having input from the teachers, and co-constructing the investigative process through the use of online resources were common practices in both classes (Solomon, 1987).

7.2.4   Creating a Place for Pūtaiao/Science Learning at the Wharekura

Pūtaiao/science was taught in the kura in English during the pre-phase and bilingually in the next two phases. Students learnt science ideas, the scientific terminology in English, and the medium of delivery was Te Reo Māori. Where appropriate, both teachers included Mātauranga Māori and students were accepting of both ways of knowing, Māori and Western science.

According to Michie, Hogue, and Rioux (2018), 'A constructivist view of science may consider alternative worldviews that include such aspects as spirituality, sharing and intuition to be incorrect understandings or misconceptions of science' (p. 1208). To consider science only through the dominant Western lens has negative consequences (Michie et al., 2018). Indigenous knowledge is socially and culturally embedded (Lederman, Abd-El-Khalick, & Schwartz, 2015) and therefore it is part of an indigenous person's culture and *being*. It can be said that when a Māori student makes sense of a science concept, their indigenous knowledge is part of the understandings they construct. It is legitimate to accept it as an alternative conception, a different way of seeing and understanding. This was evident in Rawiri's model of DNA (Fig. 6.2). He had visualised DNA as a tree, which had strong roots and shoots emerging from it. He drew on his Mātauranga Māori to draw DNA, which carried all the genetic information for a particular organism. Rawiri also included whānau, whakapapa and ancestors which showed that he understood that genes are passed down from ancestors, and whānau has an important role in the genetic identity of an individual. In his understanding of inheritance, he had shown an understanding of heredity, where the genes of ancestors are passed down through the whānau to the individual. It would be difficult to say he had a misconception of heredity; it was an alternative view of constructing his own understanding. Similarly, Māia drew upon her knowledge of pātaka when asked to build a structure. These students were not only drawing upon their experiences but also showing their ability to transfer knowledge from one domain to another. Tāne eloquently described his understanding of science and Māori knowledge (Sect. 6.5); he was able to accommodate both kinds of knowledge.

There are two schools of thought about teaching science in Māori medium schools. The purists believe that as kura are underpinned by Kaupapa Māori philosophy, science that is taught in kura should be translated into Māori and taught in Māori. With this view, a dictionary of Pūtaiao (Māori Science) has been produced, which has scientific terms likely to be used in school science translated into Te Reo Māori. As we have said in Chap. 1, this has proven to be problematic for science teaching and learning in kura because wharekura are often unable to employ teachers, who are both fluent in Te Reo Māori and who have a background in science. This is a result of the decades of colonisation of Māori people and the severe reduction of native speakers, so it also adds a burden on students to learn new vocabulary. In the absence of such teachers, either the wharekura do not offer a full range of science options to their students or look for alternative ways, for example, using intensive but short science wananga. The purist way is understandable; they would like to have all tertiary science education accessible to their students in Te Reo Māori. A pragmatist view is put forth by Māori philosopher and scholar Georgina Stewart. Stewart (2017) proposes that Western science could be taught in English in the secondary school which would give an opportunity to Māori students attending wharekura to study sciences and continue their science education in English after they leave. She argues that 'The whole amount spent on education in this country would not suffice to provide a complete science lexicon in te reo Māori: it is clearly not something that can be achieved by a few contracts and curriculum projects' (p. 32). Stewart also proposes a bilingual approach to science teaching in the kura where Western science can be taught in Te Reo Māori and can include science terminology translated into Māori. In the following quote, Stewart (2017, p. 59) reminds us of the commitment of Māori students to be bilingual:

> 2.2 Mō ngā tamariki, kia rua ngā reo. Ko te reo o ngā mātua tūpuna tuatahi, ko te reo o tauiwi tuarua. Kia ōrite te pakari o ia reo, kia tū tangata ai ngā tamariki i roto i te ao Māori, i roto hoki i te ao o Tauiwi [The aim is for the children to be bilingual. First let them develop competence in Māori, secondly in English. Equal competence in both languages will allow the children to achieve their potential in the Māori world as well as in the non-Māori world]. (Te Rūnanganui o Ngā Kura Kaupapa Māori, 2008, p. 736)

If the intention is for Māori students to be bilingual with the primacy of the Māori language, then this kura is doing just that. Their students are fluent in Te Reo Māori and are learning English as well as succeeding in both the Māori and non-Māori worlds. This can be seen in the student voice throughout the book, where Year 9 and 10 students have shown competency in communicating their ideas in English.

Further discussion on this issue is beyond the scope of this book (If interested in reading more about these ideas see, McKinley & Stewart, 2012; Stewart, 2011, 2017).

7.2.5 Ako Reciprocity of Researcher Learning Through the Project

The two researchers, Azra and Craig, come from different backgrounds and it is understandable that they have learnt different things from doing research in the wharekura.

Azra's past and her new learnings

I spent the first 22 years of my life in India and was educated there. I have lived in New Zealand for twice as long and have had the privilege of teaching in early childhood, primary, intermediate and secondary schools and at university, and of completing a doctorate in science education. Relevant to this project, all my teaching has been in English-medium schools. My awareness of Māori students and their achievement in science was raised and became a concern during 10 years of teaching in an English medium secondary school. As I worked with Māori students, I found that getting to know them as people, going and watching them play rugby, joining in a game of touch, talking with them in and outside the class, organising revision and helping with lessons on the school marae, ringing their parents when *they did well* in science, and having high expectations of work and achievement, worked. At this point, I was not aware of the research, or indeed international assessments that were reporting lower engagement and underachievement of Māori students in science. Towards the end of those 10 years, I found that my Māori students respected me, and they accepted the challenge to achieve in science and to participate in out-of-class science activities such as science competitions, science fairs and attending science summer schools at which they were encouraged and supported to participate. Māori students in my classes succeeded. While working at Wellington College of Education in 2001, I interviewed a group of ten students, both Māori and others (from my previous school), to find out their views about science learning. One statement made by a Māori student stands out; he said 'Miss you *really* know us, *care* about us, *expect* us to do *our best*, and when we do well you are *so* happy'. I did not know that I did any of this, but as my awareness was raised, I made a *real* effort to do so.

Since then, I have read and researched science teaching and learning and what students learn through science investigation. The findings from my colleague Robin Averill's research have validated that caring and being respectful helps to build relationships with Māori students (Averill, 2012; Averill & Clark, 2012). Much has been written about the importance of building relationships with Māori students and how powerful such relationships can be for the engagement and learning of Māori students in the New Zealand mainstream (English medium) schools (Bishop & Berryman, 2006; Bishop, Berryman, Cavanagh, & Teddy, 2009; Bishop, Berryman, Tiakiwai, & Richardson, 2003; Glynn, Cowie, Otrel-Cass, & Macfarlane, 2010). However, I really learnt about building relationships researching in the kura.

Having no experience of working in a Māori-medium school, when we proposed the Beyond Play: Learning Through Science Investigation research project, we were curious to find out what science the students were learning in Māori-medium schools, so we approached the kura and invited them to participate in the proposed research project. Subsequently, we carried out this research and the findings from the participating primary and secondary school cases have been published (Moeed & Anderson, 2018). In this book, we have reported the findings from the kura case study.

Our research kura has strong leadership. The principal is forward-looking, encourages teachers to try out new pedagogical approaches and wants students to be successful at the kura and be prepared for life beyond it. Kura teachers want to do their best for their students and work hard to support and care for them. Kura students are respectful, keen to learn and to support each other. Students know they are Māori, understand the culture, practice tikanga and are successful learners. An early lesson was that face-to-face contact is best when working in such a setting.

In my entire time researching at the kura, I have not seen a student sent out of class or a teacher raising their voice, or giving a detention. Students, like students in other New Zealand schools, occasionally play up, do not get their work completed or lag behind due to absence from school. When this happens, they work in the principal's office *where she works with them*. It is not a punishment but being given another opportunity to learn in contrast to detention in most mainstream schools where students are removed from class and often the punishment is to copy out lines. Students experience *success in their learning* and *success as Māori*.

This was my first experience of researching in a Māori-medium school. I think caring about students' science learning, being respectful, following my limited knowledge of tikanga and learning more have been helpful in building strong positive relationships with the principal, teachers and students at the kura. However, there is a lot more to this relationship. My experience is that I have been accepted as a member of the kura whānau (family). A whānau is more than my previous understanding of a family; kura have accepted me as part of their community and I feel I am a part of the *lived reality* of the kura, *I belong*. Strange as it might appear, other than my home, husband and daughter, this is the closest I have felt to being with my family that I left behind in India all those years ago. I belong because as Sue says, 'You want for our students what we want for them', which is indeed the case.

Craig's past and his new learnings
As a Māori person who tries to enable his children with a Māori way of being, the importance and awareness of Te Ao Māori have always been with me. I have not, however, realised that some aspects of our tikanga (way of doing) influence our Māori students' teaching and learning environment.

On reflecting on my own schooling, a complete absence of anything Māori was a reality of a young Māori person who *was* different from other students. I was too young to realise that this difference was being Māori itself. To succeed in science, I was tacitly taught to denounce my taha Māori (Māori side) and to focus on fitting in and being the same as everyone else (Pākehā). For a while, this worked, and I became successful at science, gaining a PhD in physics. The sacrifice to this journey was a realisation that I was not successful at all and could not stand on my marae and speak my language. I have spent most of my adult life catching up with my unbalanced education and trying to give teachers an appreciation of Te Ao Māori so that young Māori are better catered for in our mainstream education.

Thankfully, Māori students have other options in education nowadays and parents/teachers are more aware of the consequences of a monocultural environment. The opportunity to work with wharekura was not new to me as I had worked as a science teacher also teaching in English within a special project. I wanted to be involved in a project in which a science programme could be implemented in Māori immersion that could acknowledge a Māori worldview, weave Western and Māori science together, be sustainable, and that other communities could adopt. What are the elements of a successful programme?

As researchers entering into a space that is governed by Māori kawa (rules), an appropriate way of initialising the project is needed. The credibility, trust and potential working relationship of kura with tauiwi (foreigners to the kura) is at stake and Māori tikanga is needed to engage with kura. I had connections with kura staff, who could vouch for us to at least discuss our research ideas. This brokerage was made easier with my knowledge of the expected tikanga that solidified the working relationship. A broker in other communities might also be beneficial to other projects.

7.3   A Professional Development Model for Teaching and Learning Pūtaiao/Science in a Kura

In this section, we present our model for developing a programme for Pūtaiao/science in the research kura. We unpack the model in Fig. 7.2. Our analysis has shown that the following factors have led to the success of this model:

- *The principal and kura whānau allowing science teaching in Years 9 and above in English.*
- *Sue and Liz, being Māori and having sound pedagogical practices, knowing the students and their willingness to learn science for their students.*
- *Sue's leadership in supporting all kura teachers in primary school to introduce Pūtaiao teaching.*
- *Locating the professional development in kura.*
- *Role of experts and researchers in starting and sustaining a programme.*
- *Funding.*
- *Respectful relationships between the kura whānau and researchers.*

The big question is, can this be replicated in other kura? We believe, it can.

As noted earlier in this book, the principal and the whānau were cognisant that their children were aware of their Māori identity and they had fluency in Te Reo Māori and some of the aspects of *being* Māori. Now they wanted their students to have the opportunity to learn science as well. Yes, the schools are underpinned by the Kuapapa Māori philosophy of teaching in Te Reo Māori but they were comfortable with their students being taught in English, a compromise that many kura are having to decide given the lack of expertise of teachers in both science and Te Reo Māori. The two teachers had pedagogical approaches that worked for them and their students, and they knew the students and context very well. Sue is a strong leader who knows the curriculum and its requirements. This came across when she supported teachers to select a context, wai (water) in the first year (Material World), life cycles of butterflies and swan plants (Living World) in the following year, and she has decided on a Physical World context for the year after.

In most professional development programmes, teachers go away from their school to a different site to do their professional development. In the pre-phase, the researchers swapped their roles between one being an *expert*, and the other, the *researcher*. The content was responsive to the learning needs of the teachers. For example, teachers said they did not know any physics or chemistry, so the 2 h sessions at the beginning were in these areas. The professional development was done by negotiating the date and time with the teachers so they were not required to attend the professional development when they had reports to write, or kapahaka competitions

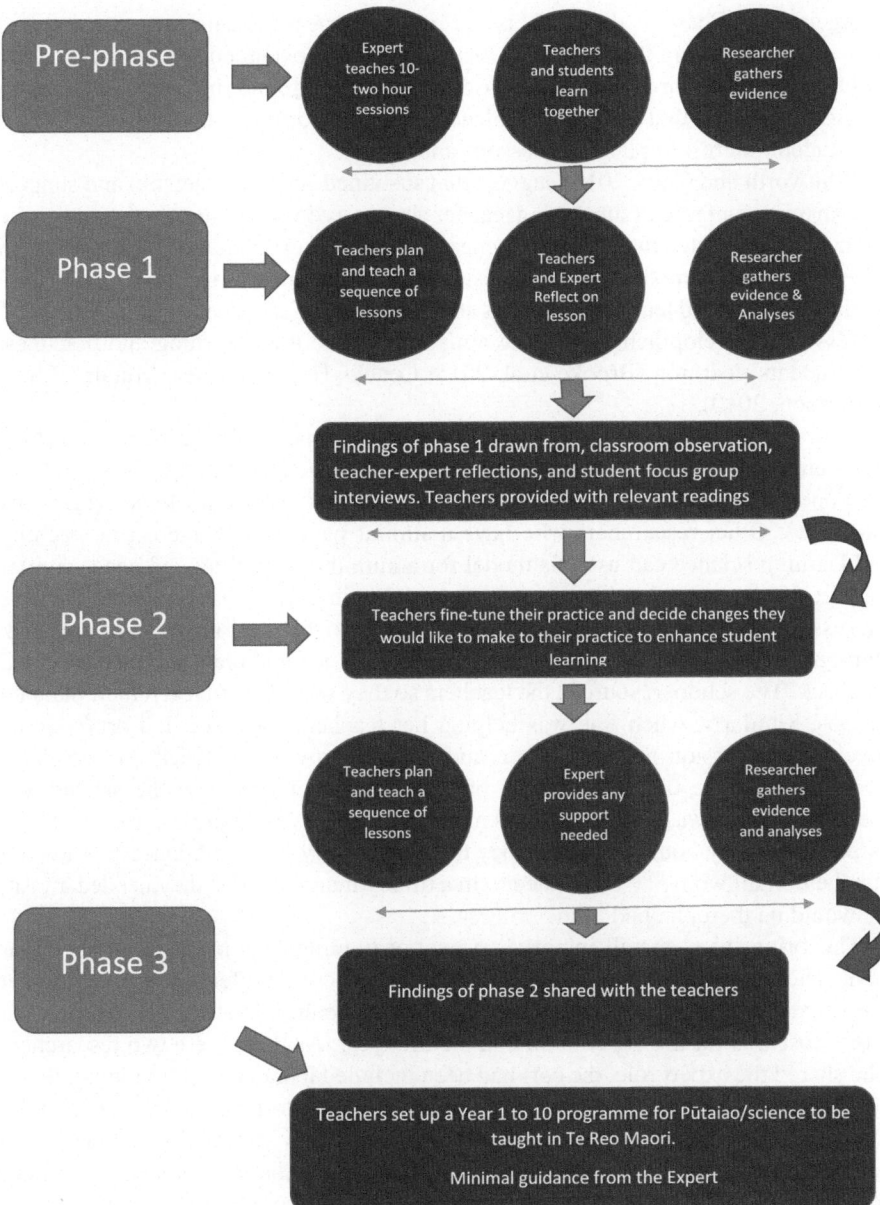

Fig. 7.2 Developing a Pūtaiao/Science programme at a kura

to organise. In Phase 1, the teachers did not just repeat the same content that had been delivered during the pre-phase; they planned and extended student learning by building on their own and students' prior learning. We found this very encouraging because it showed that as their confidence grew, they drew on their other strengths and student interest to plan their lessons and units.

Whitworth and Chiu (2015) suggest that sustained school leadership and support play an important role in supporting teacher change as do contextual factors and these ought to be integrated into science teacher development programmes, as is the case in this model. The purpose of this professional development was to improve teachers' science learning and teaching practices and also students' science learning. Teachers' professional development in science aims to improve their learning and practices, and students' learning (Brooke et al. 2015; Loucks-Horsley, Stiles, Mundry, Love, & Hewson, 2010).

Personal interest of the experts and researchers helped to build the relationships that were collectively focused on 'we want our students to learn science and have the opportunity to continue with science should they choose to do so' (Principal interview). Other researchers who have a similar passion and interest in teaching and learning science can use this model for a kura that is willing and needs similar support. It may appear a time-consuming task but in our experience, other than the ten sessions, the rest of the time needed for professional development was minimal. During Phase 1 we met for talking through the suggested unit plan, and then provided feedback. The school resourced the teachers so they were able to teach their planned lessons. Similarly, when Sue was helping her teachers we had a 1 h professional development session to support her, and later to provide feedback on her plans. Looking back, the time we gave to professional development in the second and subsequent years was minimal. It was more, the teachers knowing that they had help on hand that they could ask for if they needed it. They had confidence in knowing that their email would be responded to in a timely manner and if they needed a visit, we would be there, helped in the process.

This brings us to the all-important aspect of funding such professional development. The cost of funding is greatly reduced because the professional development was situated at the kura, which reduced the payment required for teacher relief. So the actual cost was for the expert's time in the first year. As there were two researchers who shared the expert role, the cost had been included in the research funding. Some of the funding for teacher release was used by the kura to fund all teachers and the students in the primary school to spend a morning at the Marine Education Centre. Teacher enthusiasm for this was high and the teacher in charge of the middle school approached a local bus company to pay for the bus. We could argue that this did not have anything to do with the project, but the trip only took place because the teacher was developing a unit on wai. What was really interesting was that all students were asked to have a question about something they wanted to find out and most came back with the answer. Unfortunately, the student who had a question about octopus did not get it answered even though he saw them. His reasons, 'Whaea, there were so many questions to ask! I forgot that one!'

The teachers then used the trip to draw upon students' experience to extend their learning beyond the local stream.

In sum, the key factors in the success of the model were trust, respect, acknowledgement and autonomy.

References

Abrahams, I., & Millar, R. (2008). Does practical work really work? A study of the effectiveness of practical work as a teaching and learning method in school science. *International Journal of Science Education, 30*(14), 1945–1969. https://doi.org/10.1080/09500690701749305.

Abu-Tineh, A. M., & Sadiq, H. M. (2018). Characteristics and models of effective professional development: The case of school teachers in Qatar. *Professional Development in Education, 44*(2), 311–322. https://doi.org/10.1080/19415257.2017.1306788.

Averill, R. (2012). Caring teaching practices in multiethnic mathematics classrooms: Attending to health and well-being. *Mathematics Education Research Journal, 24*(2), 105–128. https://doi.org/10.1007/s13394-011-0028-x.

Averill, R., & Clark, M. (2012). Respect in teaching and learning mathematics: Professionals who know, listen to and work with students. *Set: Research Information for Teachers, 3*, 50.

Bishop, R., & Berryman, M. (2006). *Culture speaks: Cultural relationships and classroom learning*. Wellington: Huia.

Bishop, R., Berryman, M., Tiakiwai, S., & Richardson, C. (2003). *Te Kotahitanga: The experiences of year 9 and 10 Maori students in mainstream classrooms*. Wellington: Ministry of Education.

Bishop, R., Berryman, M., Cavanagh, T., & Teddy, L. (2009). Te kotahitanga: Addressing educational disparities facing Māori students in New Zealand. *Teaching and Teacher Education, 25*(5), 734–742. https://doi.org/10.1016/j.tate.2009.01.009.

Glynn, T., Cowie, B., Otrel-Cass, K., & Macfarlane, A. (2010). Culturally responsive pedagogy: Connecting New Zealand teachers of science with their Māori students. *The Australian Journal of Indigenous Education, 39*(1), 118–127. https://doi.org/10.1375/s1326011100000971.

Goldsworthy, A., Watson, R., & Wood-Robinson, V. (1998). Sometimes it's not fair. *Primary Science Review, 53*, 15–17.

Hodson, D. (1990). A critical look at practical work in school science. *School Science Review, 70*(256), 33–40.

Lederman, N. G., Abd-El-Khalick, F., & Schwartz, R. (2015). Measurement of NOS. In R. Gunstone (Ed.), *Encyclopedia of science education* (pp. 704–708). Dordrecht: Springer.

Loucks-Horsley, S., Stiles, K. E., Mundry, S., Love, N., & Hewson, P. W. (2010). *Designing professional development for teachers of science and mathematics* (3rd ed.). Thousand Oaks, CA: Corwin Press.

Loughran, J., & Berry, A. (2011). Making a case for improving practice: What can be learned about high-quality science teaching from teacher-produced cases? In D. Corrigan, J. Dillon, & R. Gunstone (Eds.), *The professional knowledge base of science teaching* (pp. 65–81). Dordrecht: Springer.

McKinley, E., & Stewart, G. (2012). Out of place: Indigenous knowledge in the science curriculum. In B. J. Fraser et al. (Eds.), *Second international handbook of science education. Springer international handbooks of education* (Vol. 24, pp. 541–554). Dordrecht: Springer. https://doi.org/10.1007/978-1-4020-9041-7_37.

Michie, M., Hogue, M., & Rioux, J. (2018). The application of both-ways and two-eyed seeing pedagogy: Reflections on engaging and teaching science to post-secondary indigenous students. *Research in Science Education, 48*(6), 1205–1220.

Millar, R. (2010). *Analysing practical science activities to assess and improve their effectiveness.* Hatfield: Association for Science Education.

Millar, R., Tiberghien, A., & Le Maréchal, J. F. (2002). Varieties of labwork: A way of profiling labwork tasks. In D. Psillos & H. Niedderer (Eds.), *Teaching and learning in the science laboratory* (pp. 9–20). Dordrecht: Kluwer Academic.

Ministry of Education. (2007). *The New Zealand curriculum.* Wellington: Learning Media.

Moeed, A., & Anderson, D. (2018). *Learning through school science investigation: Teachers putting research into practice.* Singapore: Springer.

Murray, S. (2007). *Achievement at Maori immersion and bilingual schools: Update for 2005 results.* Wellington, NZ: Demographic and Statistical Analysis Unit (DSAU), Ministry of Education.

Nui, Te Rūnanga, & Office, Education Review. (2008). *A framework for review and evaluation in Te Aho Matua Kura Kaupapa Māori.* Wellington, NZ: Education Review Office.

Osborne, J. (2015). Practical work in science: Misunderstood and badly used? *School Science Review, 96,* 357.

Rofe, C., Moeed, A., Anderson, D., & Bartholomew, R. (2016). Science in an indigenous school: Insight into teacher beliefs about science inquiry and their development as science teachers. *The Australian Journal of Indigenous Education, 45*(1), 2015, 91–99. https://doi.org/10.1017/jie.2015.32.

Solomon, J. (1987). Social influences on the construction of pupils' understanding of science. *Studies in Science Education, 14*(1), 63–82.

Stewart, G. (2011). Science in the Māori medium curriculum: Assessment of policy outcomes in Pūtaiao education. *Educational Philosophy and Theory, 43*(7), 724–741. https://doi.org/10.1111/j.1469-5812.2009.00557.x.

Stewart, G. (2017). A Māori crisis in science education? *New Zealand Journal of Teachers' Work, 14*(1), 21–39. doi.org/https://doi.org/10.24135/teacherswork.v14i1.101

Whitworth, B. A., & Chiu, J. L. (2015). Professional development and teacher change: The missing leadership link. *Journal of Science Teacher Education, 26*(2), 121–137. https://doi.org/10.1007/s10972-014-9411-2.